Advances in Information Systems
and Management Science

Band 37

T0135866

Advances in Information Systems and Management Science

Band 37

Herausgegeben von

Prof. Dr. Jörg Becker
Prof. Dr. Heinz Lothar Grob
Prof. Dr. Stefan Klein
Prof. Dr. Herbert Kuchen
Prof. Dr. Ulrich Müller-Funk
Prof. Dr. Gottfried Vossen

Stephan Dlugosz

Multi-layer Perceptron Networks for Ordinal Data Analysis

Order Independent Online Learning by Sequential Estimation

Logos Verlag Berlin

λογος

Advances in Information Systems and Management Science

Herausgegeben von

Prof. Dr. Jörg Becker, Prof. Dr. Heinz Lothar Grob,
Prof. Dr. Stefan Klein, Prof. Dr. Herbert Kuchen,
Prof. Dr. Ulrich Müller-Funk, Prof. Dr. Gottfried Vossen.

Westfälische Wilhelms-Universität Münster
Institut für Wirtschaftsinformatik
Leonardo-Campus 3
D-48149 Münster

Tel.:   +49 (0)251 / 83 - 3 81 00
Fax: +49 (0)251 / 83 - 3 81 09
http://www.wi.uni-muenster.de

Bibliografische Information der Deutschen Nationalbibliothek

Die Deutsche Nationalbibliothek verzeichnet diese Publikation in der
Deutschen Nationalbibliografie; detaillierte bibliografische Daten sind
im Internet über http://dnb.d-nb.de abrufbar.

ISBN 978-3-8325-1984-1
ISSN 1611-3101

D 6 2008

Logos Verlag Berlin GmbH
Comeniushof, Gubener Str. 47
10243 Berlin

Tel.:   +49 (0)30 / 42 85 10 90
Fax: +49 (0)30 / 42 85 10 92
http://www.logos-verlag.de

Preface

Data Mining deals with well-established statistical problems like clustering and classification—but in the context of large and heterogeneous data sets. Such a database poses problems from various points of view: statistics, numerics and computer science. For that reason, researchers from those fields as well as from artificial intelligence have contributed to data mining from their specific perspective. The solutions proposed, however, quite often suffer from serious drawbacks looked upon from one of the other points of view. For instance, computer scientists tend to come up with procedures processing the data sequentially in a way that renders the solution order-dependent. In other words, the forecast etc. depends on the identification number of the objects under consideration. That feature, however, is completely unacceptable from a statistical point of view. The statistical remedies to the problem (batch-processing, averaging) in turn, lead to numerical complications that are hard to overcome. Another aspect, on which computer scientists and statisticians quite often disagree, is the numerical handling of data, typically represented in form of numbers. The idea of measurement-scales is hard to convey to computer scientists that deal with all kind of variables and attributes in a quantitative fashion. That again makes it hard to give any statistical meaning to the results. Error backpropagation networks (EBPN) provide an excellent class of examples, where both of the above mentioned controversies arise.

Stephan Dlugosz made them the basis of his thesis. He found a very encouraging tack to the order problem and gave fresh impetus to the treatment of ordinal variates.

ULRICH MÜLLER-FUNK

Münster
July 2008

Contents

List of Tables

List of Figures

Abbreviations

ADALINE	**Ada**ptive **Li**near **N**euron
ART	**A**daptive **R**esonance **T**heory
a.s.	**a**lmost **s**ure
BLUE	**B**est **L**inear **U**nbiased **E**stimator
CDF	**C**umulative **D**istribution **F**unction
CART	**C**lassification **a**nd **R**egression **T**ree
CT	**C**lassification **T**ree
EBPN	**E**rror **B**ack-**p**ropagation **N**etwork
EM	**E**xpectation **M**aximization
FFNN	**F**eed-forward **N**eural **N**etwork
GLM	**G**eneralized **L**inear **M**odel
GLIM	**G**eneralized **L**inear **I**nteractive **M**odeling
i.i.d.	**i**ndependent **i**dentically **d**istributed
IRLS	**I**teratively **R**e-weighted **L**east **S**quares
LDA	**L**inear **D**iscriminant **A**nalysis
LM	**L**agrange **M**ultiplier (test)
LR	**L**ikelihood **R**atio
LS	**L**east **S**quares
MDS	**M**ulti-**D**imensional **S**caling
MISD	**M**ultiple **I**nstruction **S**ingle **D**ata
ML	**M**aximum **L**ikelihood
MLP	**M**ulti-**l**ayer **P**erceptron (network)
MSE	**M**ean **S**quared **E**rror

OBD	**O**ptimal **B**rain **D**amage
OBS	**O**ptimal **B**rain **S**urgeon
PCA	**P**rincipal **C**omponents **A**nalysis
PDF	**P**robability **D**ensity **F**unction
PPR	**P**rojection **P**ursuit **R**egression
RT	**R**egression **T**ree
S-CART	**S**tructured **R**egression **T**ree and **C**lassification **T**ree
SVM	**S**upport **V**ector **M**achine
VC	**V**apnik-**C**hernovenkis (dimension or theory)

Symbols

$\mathbb{1}_M(x)$	indicator function with value one if $x \in M$ and zero else
a_i	summed input of node i
$\boldsymbol{\beta}$	parameter vector in linear models
β_0	constant parameter in linear models
$\boldsymbol{\beta}^-$	parameter vector in linear models without constant
β_{0r}	cut point between class r and $r+1$ for an ordinal variable with an underlying metric variable in linear models
$\mathfrak{B}^k$	k-dimensional Borel set
$\mathcal{C}(\mathcal{X})$	continuous functions on compact $\mathcal{X}$
$\mathcal{C}^m(\mathcal{X})$	m-times continuous differentiable functions on $\mathcal{X}$
$\mathrm{Cov}(X,Y)$	covariance of two random variables $X, Y \in \mathbb{R}$
$D(\cdot)$	domain of $\cdot$
d	dimension of the input vector
δ_j	error associated to node j in a neural network during backpropagation
$\mathbb{E}$	expectation value
ϵ	error term
η	learning rate/step size in numerical optimization
$f(\cdot)$	regression function
$F(\cdot)$	arbitrary cumulative distribution function
$g_i(\cdot)$	transfer function of node i
k	dimension of the output vector in neural networks, number of distinct classes in classification
$\mathrm{logit}(\cdot)$	logit function (cf. Table 2.4)

$\ell(\cdot)$ transformation functions (used to make model independent to the choice of the scale, see 2.1.3)

n size of the sample

$P(\omega)$ probability, that element ω from a countable set Ω is chosen

$P(\omega|A)$ conditional probability, that ω is chosen from Ω with knowing, that only elements from $A \subset \Omega$ have positive probability of being chosen

r index of the category

R^2 coefficient of determination

θ_r cut point between class r and $r+1$ for an ordinal variable with an underlying metric variable

τ coefficients in Taylor series expansions

$\mathrm{Var}(X)$ variance of a random vector $X \in \mathbb{R}$

w_{ij} weight of the connection between nodes i and j

$\mathbf{X}$ design matrix in linear models

Y response variable

$\tilde{Y}$ (in ordinal models) metric underlying variable for ordinal response variable Y

z_i output of node i

$\langle \cdot, \cdot \rangle$ scalar product

$\boldsymbol{x}^t$ transposed vector

$z_i^{(n)}$ value of node i of a neural network for pattern n during backpropagation

$\mathrm{diag}(\boldsymbol{v})$ diagonal matrix formed by the entries of vector v

$\lambda\!\!\!\lambda$ Lebesgue measure

$\mathcal{L}^p(\mathcal{X})$ p-th power Lebesgue integrable functions on $\mathcal{X}$

$\nabla f(\boldsymbol{x})$ gradient of f at $\boldsymbol{x}$

$\nabla^2 f(\boldsymbol{x})$ Hessian of f at $\boldsymbol{x}$

$(\boldsymbol{x}_n, \boldsymbol{y}_n)_{n \in \mathbb{N}}$ sample, consisting of n tuples $(\boldsymbol{x},\boldsymbol{y})$ with $\boldsymbol{x} \in \mathbb{R}^d$ and $\boldsymbol{y} \in \mathbb{R}^k$

i, j indexes used in neural network models

$\iota,\ \kappa$ and λ indexes used in Taylor series expansions

$\Phi(\cdot)$ Gaussian Cumulative Distribution Function

$\sum_{i\to j} x_{i,j}$ $\sum_i x_{i,j}$ for $j \in J, i \in I$, I, J index sets

$\|\cdot\|$ supremum norm

1 Introduction

Surveys are a popular quantitative research method in the social sciences. Many of the variables used in surveys are measured on an ordinal scale. This is the reason why analysis of ordinal data is an important topic especially in the social sciences, maybe even more important than the analysis of metric data.

However, the analysis of ordinal data has not been well developed in theory, yet. This is one of the reasons why in psychological, medical or social studies data (even measured on an ordinal scale with "check-boxes") are often supposed to be metric (Gautam, Kimeldorf and Sampson, 1996). A metric scale is often assumed without discussing the underlying assumptions concerning this measurement of scale (Woodward, Hunter and Kadlec, 2002; Taylor, West and Aiken, 2006). However, there are some good parametric (O'Connell, 2006; Liu and Agresti, 2005) and nonparametric (Kramer, Widmer, Pfahringer and de Groeve, 2000, 2001) models for ordinal regression, whereas the later ones are often unknown to the scientific community. The usefulness of ordinal regression in contrast to ordinary categorical regression is theoretically plausible as the given order information is used in addition. Comparing ordinal regression to ordinary nominal categorical regression in concrete data analysis confirms this expectation (Campbell and Donner, 1989; Taylor and Becker, 1998; Norris, Ghali, Saunders, Brant, Galbraith, Faris and Knudtson, 2006).

Linear models are very popular models of explanation in science because their parameter estimates are easy to interpret. Yet linear models are often unsuitable for use in practical applications, especially for forecasting. Instead, nonlinear, nonparametric or semiparametric models are used. Among these, feed-forward neural networks are very popular as they are very flexible both in terms of model complexity and ability to deal with metric and categorical data on input and output side.

Unfortunately, nonlinear semiparametric models for ordinal data are rare. In economic applications, in which ordinal alternatives such as customer credit scoring or company ratings play a major role, ordinal feed-forward neural networks can be used to obtain better model fits and—as follow-ups—higher reliabilities of prognoses, for instance.

1.1 Ordinal Data in Neural Networks

The most important[1] nonlinear regression techniques are multi-layered feed-forward neural networks with parameter estimation via backpropagation. This kind of model has been shown to provide a very flexible model structure that allows general function representation for (asymptotically) every interesting function (cf. section 4.1). Although neural networks are statistical tools, they have been developed by the artificial intelligence research community. That is the reason why neural networks are described and classified in taxonomies according to their algorithmic nature instead of their (statistical) model or purpose.

Feed-forward neural networks with backpropagation are (among others[2]) the first neural network techniques analyzed and categorized under a statistical point of view (White, 1989b; Amari, 1990; Ripley, 1997). The "learning" algorithm is the most interesting "part" of the model for statisticians, because of the statistical properties of the provided estimators. Both, offline and batch learning use the Least Squares (LS) estimation principle as optimization criterion. The online version is a special form of the Robbins-Monro algorithm (Robbins and Monro, 1951). This algorithm has been shown to provide an approximate bias-free LS estimator (White, 1989c). One of its major drawbacks is the use of a non-sufficient statistic, resulting in dependence on the order in which the data occurs in a finite sample.[3] This is not a negligible effect as example 1.1 demonstrates.

---

[1]i.e. "applied to many practical tasks of data analysis", especially economical applications (Kuan and White, 1994).

[2]i.e. Kohonen Networks or the simple perceptron.

[3]This problem is still unresolved (Sarle, 1997b). Bishop (1991) addresses a similar problem.

Example 1.1

Assume a linear regression model with the regression function $f(x) = 2x + 1$. A small simulation of standard Gaussian additive error variables yields the following sample (rounded):

x	1	10	2	4	7	8	6	5	9	3
y	0.7	19.45	4.62	8.8	16.77	15.81	13.45	8.15	20.25	10.83

Tab. 1.1: Regression example

These data have been used for estimating the parameters of the simple linear regression function $y = ax + b$, $a, b \in \mathbb{R}$ by the backpropagation algorithm[4]: $b_{n+1} = b_n - \eta_n(a_n x_n + b_n - y_n)$ and $a_{n+1} = a_n - \eta_n x_n(a_n x_n + b_n - y_n)$. Starting point was $(a_0, b_0) = (0, 0)$ with decreasing learning parameter $\eta_n = 0.1 \cdot 1/n$ (Bishop, 1995, p. 264). Using the ordering of the data shown in Table 1.1, the resulting parameter vector was $(2.44, -0.61)$; using the reverse ordering yields $(1.77, 1.15)$. Using the same data in offline learning after 5, 10 and 20 steps the algorithm estimates the parameters with $(2.08, 0.32)$, $(2.07, 0.32)$ and $(2.07, 0.33)$ respectively[5]. The sorted data set by ascending X values results in even worse $((1.87, 0.69)$ vs. $(0.74, -0.03))$ parameter estimates. Although this ordering is unlikely, it cannot be avoided in online learning. For comparison, the standard LS estimation for this data is $(2.03, 0.68)$.[6]

Obviously, the offline learning gradient descent minimization gives the best result; even after only 5 iterations. The difference between the parameter estimation for offline learning and the exact LS estimator is a result of the low convergence rate of the gradient descent algorithm. The latter one has been used to make sure that the differences only depend on the online and offline principles, not on the approximation vs. exact calculation. There is a great variety of local and global optimization techniques for offline learning (Bishop, 1995), but only a few local optimization techniques have been adopted for online learning.

The example demonstrates clearly that the ordering of the data set may play an important role in estimation. Confronted with parameter estimation

---

[4]Further details on this estimation technique are provided in section 4.2.

[5]The learning parameter had been reduced by a factor of 10 to compensate the summation of the various errors used in the batch learning principle.

[6]Deviations from the parameters used in the simulation are a result of the small sample size and the relatively high variation of the error term.

for statistical models that are order-dependent, statisticians usually try to calculate the appropriate average for the different results, i.e. they look for a collection of parameter estimators, generated by different orderings of the sample and calculate the required "order-independent" estimator as an average value of the different order-dependent estimators. However, there are some severe problems:

- Large number of permutations: How many (and which) orderings are sufficient for estimation?

- Parameter-symmetry in feed-forward neural network models for model averaging: Which are the "same" parameters that have to be averaged?

There are probably more problems that have to be solved in order to use this approach. Therefore, an alternative solution to the problem of order-dependent models—using sufficient statistics—will be proposed in this thesis.

Like linear models, feed-forward neural networks have been proposed for classification tasks. Maximum likelihood is the estimation technique in generalized linear models for classification, but feed-forward neural networks for classification are usually "trained" with a LS estimator (Duda, Hart and Stork, 2000; Bishop, 1995), thus ignoring the known distribution family. This approach also suffers from the following shortcomings concerning the network design for k distinct classes:

1) One output node with k distinct possible values (e.g. $\{1 \ldots k\}$).
   In this case, an underlying ordinal structure of the categories is required that usually does not exist. In addition, the implementation of the rounding procedure is problematic.

2) k binary output nodes indicating the chosen category.
   The problem with this design is to ensure that only one node is set at the same time. And the cross-links, that could be used to overcome this problem, would destroy the feed-forward structure of the network, which is fundamental for the backpropagation algorithm.

3) k nodes, one for every class with values in $[0, 1]$, indicating the likelihood of the corresponding class.
   This design allows a kind of affinity measure. Nevertheless, the "real" values

of affinity to a class are not known, only the class itself. Thus the usage of the Mean Squared Error (MSE) function is not appropriate.[7]

As MLP networks are used for classification and regression, they are a good choice for ordinal regression, because ordinal regression is classification with an additional order structure on the different classes but without any distance information like in regression. There are two different approaches for constructing ordinal MLP neural networks. Mathieson (1996) extends ordinary linear regression models for nominal data by using the cumulative proportional odds model by McCullagh (1980) to neural networks.[8] Da Costa and Cardoso (2005) use the "implicit" ordering of the binomial probability function. With a fixed number of tries, they construct a neural network regression onto the interval $[0, 1]$ interpreted as the probability of success p in the binomial model. Thus, they obtain an unimodal a-posteriori distribution.

These different approaches are discussed in broad in section 4.6.

1.2 Organization of this Thesis

The organization of this thesis is illustrated in Figure 1.1. Directed arrows indicate dependencies between chapters.

Chapter 2 presents some well-known foundations and definitions related to the theory of scales of measurements (cf. section 2.1), statistical and machine learning (cf. section 2.2), modeling of input-output relations (cf. section 2.3) and neural networks (cf. section 2.5). Even readers with previous knowledge should take a look at sections 2.1 and 2.5.2, where some less common aspects of these topics are discussed. A new, special type of ordinal data is introduced in subsection 2.1.4.

Modeling approaches for ordinal data in input-output models are introduced in chapter 3. There are some innovative aspects of ordinal explanatory variables presented in section 3.1. General approaches to ordinal output variables are addressed in section 3.2. In sections 3.3 and 3.4 and in subsection 3.2.3 some popular methods for ordinal regression are explained and compared.

---

[7]Using well-known error functions from GLMs for MLP networks solves this problem (cf. 4.5).

[8]Cheng (2007) describes the same idea without referring to Mathieson (1996).

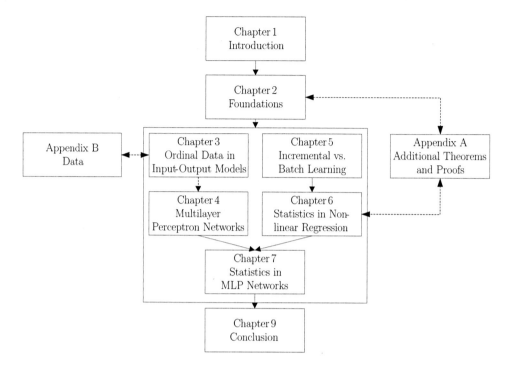

Fig. 1.1: Organization of the thesis

Foundations of multi-layer perceptron networks with their representation capabilities and the backpropagation learning algorithm are stated in chapter 4.[9] The problems of (ordinal) classification with feed-forward neural networks are introduced in section 4.5 and discussed in detail in section 4.6.

In chapter 5, statistically relevant differences and practical problems of sequential and batch learning are introduced. Order-dependence of the sequential learning principle is elaborated upon. A variety of possible solutions is presented in section 5.2.

Chapter 6 contains the definitions of infinite sufficient statistics and the approximate sufficient statistic with their use in simple (non-)linear regression with exponential nonlinearities. This is the most general part of the new approach introduced in this thesis and may be called nonlinear generalized models.

---

[9]Further readings on multi-layer perceptron networks: Bishop (1995), Ripley (1996), Haykin (1999), Duda, Hart and Stork (2000) and others.

These general results are used in chapter 7, where the sufficient statistic for parameter estimation in a multi-layer perceptron network is introduced. This is the more "practical" part of this thesis, where the new theoretical results from chapter 6 are used in a feed-forward neural network setting. Besides simple application of the results from chapter 6, the focus in chapter 7 lies on an efficient implementation of algorithms for regression (metric and ordinal) and classification, including an approach for parallel calculation.

The conclusion and an appendix with some additional theorems complete the thesis.

2 Foundations

This chapter introduces the basic terms and definitions used throughout this thesis. Many of them are well known, but they are needed to clarify the interpretation of certain terms, because they are used differently throughout the literature (especially the terms and concepts in section 2.2).

There are also some rather new aspects on measurements of scales, particularly the formal definitions of the scales (subsection 2.1.1), their influence on data analysis (subsection 2.1.3) and additional properties of scales between interval and ordinal scales (2.1.4). Basic ideas of neural networks and their relation to ordinary statistical analysis are discussed in sections 2.5 and 2.5.2.

2.1 Measurements of Scale

Stevens (1946) published an article about measurements of scale to found a basis for interpretation of statistical functionals (Stevens, 1951). While most of the statisticians[10] saw it as an important link between the empirical and the theoretical points of view (Townsend and Ashby, 1984; Knapp, 1990; Cliff, 1992, 1996a,b), many empirical researchers in the social sciences did not accept the implications associated with the set of admissible transformations (Baker, Hardyck and Petrinovich, 1966). In particular the restrictions of statistical moments and methods (Stevens, 1951, 1955) have been criticized (Lord, 1953; Burke, 1953; Anderson, 1961; McHugh, 1963). During the following years, a discussion took place on this topic and Stevens was forced to precise—but not revise—his statements (Stevens, 1959, 1968).[11] Although the concept of scales of measurement is nowadays widely accepted, there are new criticism formulated by researchers from computer science, e.g. data mining researchers (Velleman

---

[10]cf. Krantz, Luce, Suppes and Tversky (1971).

[11]Khurshid and Sahai (1993) summarize the different theories of measurement and criticize Stevens's approach as well as a broad bibliography on that topic.

and Wilkinson, 1993). Besides the criticism related to the implications of the
concept of measurement scales, the scales can also be criticized (cf. 2.1.2).

2.1.1 Definitions

Definition 2.1 (nominal scale)
*Let $(M, \sim)$ be an empirical relative on the set M with $\sim \subset M \times M$ satisfying
the following properties:*

(i) $\forall_{a \in M} a \sim a$ (reflexivity)

(ii) $\forall_{a,b \in M} a \sim b \Leftrightarrow b \sim a$ (symmetry)

(iii) $\forall_{a,b,c \in M} a \sim b \wedge b \sim c \Rightarrow a \sim c$ (transitivity)

*A homomorphism (ψ) between $(M, \sim)$ and the numerical relative $(\mathbb{R}, =)$ with
$= \subset \mathbb{R} \times \mathbb{R}$ and $a \sim b \Leftrightarrow \psi(a) = \psi(b)$ is called a* nominal scale.

There exists an equivalence class of numerical relatives for every nominal
empirical relative, which is based on injective functions. Therefore, the only
meaningful[12] statistics are those which are invariant to this class of functions.[13]

Definition 2.2 (ordinal scale)
*Let $(M, \sim, \succeq)$, $\succeq \in M \times M$ be an empirical relative on the set M with $\sim$ being
reflexive, symmetric and transitive (see (i), (ii) and (iii) of Definition 2.1). $\succeq$
should be:*

(iv) $\forall_{a,b,c \in M} a \succeq b \wedge b \succeq c \Rightarrow a \succeq c$ (transitivity)

(v) $\forall_{a,b \in M} a \succeq b \vee b \succeq a$ (totality)

*With $\forall_{a,b \in M} a \sim b \Leftrightarrow a \succeq b \wedge b \succeq a$ the relation $\succeq$ is sufficient. A homomorphism
(ψ) between $(M, \succeq)[= (M, \sim, \succeq)]$ and the numerical relative $(\mathbb{R}, \geq)$ with $a \succeq
b \Leftrightarrow \psi(a) \geq \psi(b)$ is called an* ordinal scale.

There exists an equivalence class of numerical relatives for every ordinal
empirical relative, which is based on strictly isotone functions. Meaningful
statistics therefore has to be invariant under strictly isotone transformations.

---

[12]The topic meaningful statistics is amongst others discussed in (Luce, 1959), (Michell,
1986) and (Hand, 1996).

[13]Proofs for this and the other equivalence classes may be found in the appendix A.

Definition 2.3 (interval scale)

Let $(M, \succeq)$ be an empirical relative on the set M with $\succeq$ being reflexive, symmetric, transitive and total (see (i), (ii), (iv) and (v) of Definition 2.2). Also let $(M \times M, \overset{\sim}{\succeq})$ with relation $\overset{\sim}{\succeq}$ on $M \times M$ with $(a, b) \in M \times M$ be interpreted as "distance" of the values $g^{-1}(a)$ and $g^{-1}(b)$ with g describing a strictly isotone function $g : \mathbb{R} \to \mathbb{R}$. Standards for g^{-1} are the identity (interval scale) and a logarithm with an arbitrary basis (log-interval scale, (Stevens, 1951)). $\overset{\sim}{\succeq}$ should be:

(vi) $\forall_{a_1, a_2, b_1, b_2, c_1, c_2 \in M} (a_1, a_2) \overset{\sim}{\succeq} (b_1, b_2) \wedge (b_1, b_2) \overset{\sim}{\succeq} (c_1, c_2) \Rightarrow (a_1, a_2) \overset{\sim}{\succeq} (c_1, c_2)$
   (transitivity)

(vii) $\forall_{a_1, a_2, b_1, b_2 \in M} (a_1, a_2) \overset{\sim}{\succeq} (b_1, b_2) \vee (b_1, b_2) \overset{\sim}{\succeq} (a_1, a_2)$ *(connexivity)*

(viii) $\forall_{a, b, c, d \in M} (a, b) \overset{\sim}{\succeq} (c, d) \Rightarrow (d, c) \overset{\sim}{\succeq} (b, a)$ *(sign inversion)*

(ix) $\forall_{a, b, c, d, e, f \in M} (a, b) \overset{\sim}{\succeq} (d, e) \wedge (b, c) \overset{\sim}{\succeq} (e, f) \Rightarrow (a, c) \overset{\sim}{\succeq} (d, f)$ *(weak monotonicity)*

(x) $\forall_{a, b, c, d \in M} (a, b) \overset{\sim}{\succeq} (c, d) \overset{\sim}{\succeq} (a, a) \Rightarrow \exists_{d_1, d_2 \in M} (a, d_1) \sim (c, d) \sim (d_2, b)$ *(solvability)*

(xi) Every strictly bounded sequence $a_i \in M$ such that $(a_i, a_{i+1}) \sim (a_1, a_2)$ for all i and $(a_1, a_2) \not\sim (a_1, a_1)$ is finite (Archimedean axiom)

A homomorphism (ψ) between $(M, \succeq, \overset{\sim}{\succeq})$ and the numerical relative $(\mathbb{R}, \geq)$ with $a \succeq b \Leftrightarrow \psi(a) \geq \psi(b)$ and $(a, b) \overset{\sim}{\succeq} (c, d) \Leftrightarrow g^{-1}(\psi(a)) - g^{-1}(\psi(b)) \geq g^{-1}(\psi(c)) - g^{-1}(\psi(d))$ is called g^{-1}-interval scale.

The equivalence class of numerical relatives to this empirical relative is formed by all functions $f : \mathbb{R} \to \mathbb{R}$ with $f(x; a, b) = g(ag^{-1}(x) + b)$ with constants $a \in \mathbb{R}^{>0}, b \in \mathbb{R}$. For instance, this are linear functions $(f(x; a, b) = ax + b)$ for the standard interval scale and power functions $(f(x; k, n) = bx^a)$ for the log-interval scale.

Comment 2.4

Only strictly isotone functions may be used for g as g has to be invertible and the ordinal relation on the differences has to be preserved.

Example 2.5

Let $g(x) = x^3$ be an isotone function with its inverse $g^{-1}(x) = \sqrt[3]{x}$. Thus, we can define a $\sqrt[3]{x}$-interval scale where the distances of the $\sqrt[3]{x}$-transformed values preserve their ordering.

This very unusual construction of a scale may be useful only in very special cases, but the log-interval-scale with the simple interpretation of equal ratios and the (normal) interval scale (with $g = id$) with the interpretation of equal distances are both very common, and together they form the ratio scale.

Definition 2.6 (ratio scale)

Let $(M, \succeq, \overset{\sim}{\succeq}, \circ)$ be an empirical relative on the set M so that $\succeq$ has the properties (i), (ii), (iv) and (v) defined in Definitions 2.1 and 2.2. Let further $(M \times M, \overset{\sim}{\succeq})$ satisfy the properties (vi), (vii), (viii), (ix), (x) and (xi) defined in Definition 2.3. Additionally, the relation $\circ \in M \times M$ has to satisfy the properties:

(xii) $\forall_{a,b,c \in M} a \circ (b \circ c) \sim (a \circ b) \circ c$ (associativity)

(xiii) $\forall_{a,b,c \in M} a \succeq b \Rightarrow a \circ c \succeq b \circ c \wedge c \circ a \geq c \circ b$ (monotonicity)

(xiv) $\forall_{a,b \in M} \exists_{n \in \mathbb{N}} \underbrace{a \circ a \circ \ldots \circ a}_{n \text{ times}} \succeq b$ (Archimedean axiom)

Then, a homomorphism (ψ) between $(M, \succeq, \overset{\sim}{\succeq}, \circ)$ and the numerical relative $(\mathbb{R}^+, \geq, +)$ with $a \succeq b \Leftrightarrow \psi(a) \geq \psi(b)$, $(a,b) \overset{\sim}{\succeq} (c,d) \Leftrightarrow \psi(a) - \psi(b) \geq \psi(c) - \psi(d)$ and $\psi(a \circ b) = \psi(a) + \psi(b)$ is called ratio scale.

The equivalence class of functions associated with this scale is the class of linear functions through the origin ($f(x; a) = ax$ (with $a > 0$)).

As the ratio scale gives zero a special meaning and interpretation[14], this can also be done for the quantity 1 to be interpreted as a unity. This is called the absolute scale, and the equivalence class of possible numerical representations only contains the identity. The axioms are the same as for the ratio scale with an additional associative relation $\tilde{\circ} \in M \times M$ with $\forall_{a,b,c \in M, c \circ c \succeq c} a \succeq b \Rightarrow a \tilde{\circ} c \succeq b \tilde{\circ} c \wedge c \tilde{\circ} a \geq c \tilde{\circ} b$ and $\psi(a \tilde{\circ} b) = \psi(a) \cdot \psi(b)$. Although this completes the scale

---

[14]Axiom (xiv) restricts the numerical relative to the set $\mathbb{R}^+$, i.e. any numerical value associated to an existing object has to be greater than zero.

system, no special statistical procedure that would require data measured on
an absolute scale is known to the author.

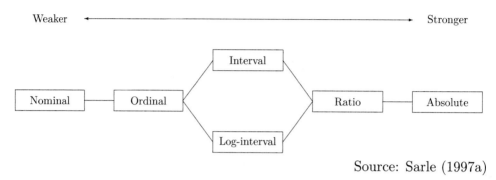

Source: Sarle (1997a)

Fig. 2.1: Overview

The nominal and the ordinal scales are called categorical. The interval,
ratio and absolute scale are usually referred to as metric scales. There are
more (uncommon) scale types, like the difference scale. A good overview is
presented in Roberts (1979) and in Narens (1985) which is more mathematically
sophisticated.

2.1.2 Criticism

Stevens' concept of measurement scales has been criticized a lot during the
first years, but finally, it was widely accepted (Khurshid and Sahai, 1993). The
criticism may be divided into two groups: First, there are concerns in accepting
the consequences that admissible transformations of scales imply to obtain
interpretable statistical results. Secondly, it is questioned, if this scale system
is complete, i.e. if every empirical measure belongs to exactly one of the scales.
While the former criticism is obsolete, the latter still is a matter of concern.

However, the old discussions retained importance as "data mining" becomes
more popular (Velleman and Wilkinson, 1993). The critics mention that it is
not clear in advance, if there either is or not a useful interpretation for the
empirical relative. The following example should illustrate the problem:
Suppose you collect data about the color of someone's hair and the risk of
getting sunburn after being a certain amount of time in the sun. The time is
obviously measured on a ratio scale; the color of the hair is nominal. However,

if the color of the hair is seen as an indicator for the skin type, you can treat this scale as ordinal with ranks red, blonde, brown and black.

In contrast to Velleman and Wilkinson, such problems with Stevens' scales are neither a problem of the scales nor of admissible scale transformations, but of treating scales assigned to data without considering the purpose of measurement. Another point of criticism is based on a misunderstanding of the term "admissible transformations".[15] This is *not* a set of allowed transformations for the data to obtain results, but a set of transformations of the measurement scale. Every methodology applied to the data should be invariant to transformations of the scale. This leads to the concept of meaningful statistics (cf. subsection 2.1.3).

There have been different approaches to solve this controversy. Michell (1986) demonstrated that the different points of view result from three different paradigms of measurement: representational (Stevens' approach), operational and classical.[16]

> The operationalist will be interested in devising operations that produce reasonably consistent numerical assignments. The representationalist will be interested in finding empirical relations that display properties similar to those of relations between numbers (e.g., orders, orders on differences, etc.) (Michell, 1986, p. 404).

Simply speaking, the operationalist is concerned with constructing useful measurements and the representationalist may use these measurements in further analysis. These different areas of responsibilities have often been mixed and caused an unfruitful discussion.

Completeness of the Scale System

Besides the general criticism, Velleman and Wilkinson present the problem of multiple mutual measurements of scale. This rises the question if the scale system by Stevens is complete and, if not, what is missing and why. While you can argue to use separated (but dependent) measurements for multiple mutual measurements like that in Velleman and Wilkinson (1993), quantities like percentages or the cardinal points are less easily resolved.

---

[15]For example, see Velleman and Wilkinson (1993, p. 68).
[16]see also Hand (1996).

Percentages can be treated as measured on an absolute scale because they are defined "scale-free" in a dimensional point of view. But we may not add up percentages, as we may leave the range $[0, 1]$. Thus, percentages may only be measured on an ordinal scale. This is questionable because information gets lost. For example the statement "20 percent of a cake is the double of 10 percent of that cake" is meaningful.

The circular structure of cardinal points prohibits even the measurement on an ordinal scale, thus only the nominal scale is left. Obviously, this will not represent the empirical relations, as differences (in terms of the shortest way around the circle) are meaningful in the empirical and numerical relative.

Another "gap" in the scales of measurement lies between the ordinal and the interval scale. There are possible scales with empirical relatives that are not interval but have a stronger structure than simple order. In section 2.1.4, we will present some possible properties defining scales of measurement between the ordinal scale and the interval scale.

2.1.3 Influence on the Analysis of Input-Output-Models

Before looking at the construction of models for data analysis of various scales, we have to ensure that these models are meaningful. The concept of meaningful statistics has been analyzed for a long time starting with Luce (1959) and ending up with Molenberghs and Verbeke (2004). The term meaningfulness will be understood as invariance of admissible transformations of the scale system in relation to decision-making (cf. Hand (1996)).

Based on the measurement of scales, Luce (1959) tried to characterize possible input-output-relations by stating that admissible transformations of the input variable should lead to admissible transformations of the output variable. This implies, for instance, that the relationship between two variables x and y on the ratio scale has to be $y = \alpha x^{\beta}$, $\alpha, \beta \in \mathbb{R}$. This rigid restriction has generated discussions, and it has been defended by Luce (1990).

In terms of meaningfulness, we have to distinguish between statistics describing properties of one variable like mean or dispersion and statistics describing relationships between different variables. In the former case, the originally noted property of invariance to scale transformations by Stevens is appropriate to define meaningful statistics, as we do not want the conclusions depend on the choice of the scale (Marcus-Roberts and Roberts, 1987). In modeling

Scaling	$D(x)$	$\ell_1(x)$	$D(y)$	$\ell_2^{-1}(z)$
nominal[a]	$\{1,\ldots,k\}$	$\begin{pmatrix} a_1 \mathbb{1}_{\{2\}}(x) + b \\ \vdots \\ a_{k-1}\mathbb{1}_{\{k\}}(x) + b \end{pmatrix}$	$\{1,\ldots,k\}$	–
ordinal[a]	$\{1,\ldots,k\}$	$\begin{pmatrix} a_1 \mathbb{1}_{\{2,\ldots,k\}}(x) + b \\ \vdots \\ a_{k-1}\mathbb{1}_{\{k\}}(x) + b \end{pmatrix}$	$\{1,\ldots,k\}$	–
interval	$\mathbb{R}$	$ax + b$	$\mathbb{R}$	$\dfrac{z-b}{a}$
ratio	$\mathbb{R}$	ax	$\mathbb{R}$	$\dfrac{z}{a}$
absolute	$\mathbb{R}$	x	$\mathbb{R}$	z

[a]There are different possibilities to $\ell_2^{-1}(z)$ covered in detail in section 3.2.

Tab. 2.1: Scale transformation functions for regression problems

relationships between different variables, we usually have parameters to be estimated. We cannot expect these parameters to be independent from the choice of the scale (as Luce (1959, 1990) does), but we can expect the functional form to be independent.

To establish a relationship between two variables, the construction of the function representing the relationship has to obey the measurements of scale. For example, given metric input $x \in \mathbb{R}^d$ and output $y \in \mathbb{R}^k$ and a function $f(\cdot)$ representing the "real" relationship between the two concepts x and y, which are measured by x and y. Each element of the input vector x has to transformed linearly in advance to applying the function f. Afterwards another linear transformation is needed to represent the measurement of the output variable y. The result is a combination of three functions ℓ_1, ℓ_2 and f:

$$\ell_1(x) = \begin{pmatrix} a_{11}x_1 + a_{10} \\ a_{21}x_2 + a_{20} \\ \vdots \\ a_{d1}x_d + a_{d0} \end{pmatrix} \qquad \ell_2(y) = \begin{pmatrix} b_{11}y_1 + b_{10} \\ b_{21}y_2 + b_{20} \\ \vdots \\ b_{d1}y_d + b_{d0} \end{pmatrix} \tag{2.1}$$

In brief, the regression function is

$$y = \ell_2^{-1} \circ f \circ \ell_1(x) + \epsilon \tag{2.2}$$

The parameters are not identified for a linear regression function f and the equation (2.2) "collapses" to the well-known simple form. The functions ℓ_1 and ℓ_2 are shown in Table 2.1 for every measurement of scale (in the one-dimensional case).

Let us have a look at a simple example:

Example 2.7 (Nonlinear Regression)
Let us assume, someone wants to analyze the relationship between two ratio scaled variables X and Y. He might have decided, that an exponential relationship is appropriate. Therefore, he tries to fit the function

$$f(x; \beta) = \exp[\beta x]$$

to an i.i.d. sample data set $(x_n, y_n)_{n \in \mathbb{N}} \in \mathbb{R}^{n \times 2}$. This functional form depends on the currently chosen scale of the output variable Y. Therefore, it is necessary to add a parameter to represent different scalings of Y:

$$f(x; \beta_1, \beta_2) = \beta_2 \exp[\beta_1 x]$$

If Y is measured on an interval scale, we need an additional parameter and obtain

$$f(x; \beta_1, \beta_2, \beta_3) = \beta_2 \exp[\beta_1 x] + \beta_3 \ .$$

Measuring X on an interval scale has got no influence on the form of the regression function, as an additional constant and β_2 could not be distinguished and the model would not be identifiable.

Therefore, the set of possible functional relationships is slightly restricted, but we will not produce meaningless results (cf. Table 2.1). The required restrictions to obtain meaningful statistical relationships are not so severe as those in Luce (1959), but sufficient.

2.1.4 Ordinal Data

As this work is focused on the analysis of ordinal data, we should have a closer look at the origins of ordinal-scaled data. Anyway, it might be interesting to search the area between ordinal and interval scale for intermediate scale types.

Sources of Ordinal Data

There are many circumstances restricting the precision in data collections. Therefore, data, measured with an interval or ratio scale, is often censored and can only be measured on an ordinal scale. Thus, we have to distinguish between simple ordinal data and censored interval or ratio scaled data, called "grouped". The main difference is the ability of being measured on an interval scale. We do not have the possibility to measure simple ordinal data on an interval scale (although it might theoretically be possible). On the contrary, "censored" data has been measured on an interval scale, but is submitted to the researcher only in an ordinal scale. This censoring may be done during data acquisition or afterwards to obey privacy laws.

Sometimes, even rounded interval scaled data[17] might be ordinally scaled. This is the case if the rounding process is not the same on the complete data range. On the other side, "interval scaled" data without open intervals at the tails of the range may be treated like interval data as long as the intervals are of equal size.[18]

Although Nelder (1990) classifies simple ordinal data into an ordinal scale with and without underlying (latent) metric variable, he does not explain any differences in data analysis or interpretation of results. Examples for simple ordinal data are grades, company ratings and levels of education.

Data is called *grouped*, if it has been interval-censored (rounded) within intervals, which are not necessarily equidistant. Data is called *truncated*, if there are values (at the lower and higher end), which collect all values lying at the lower or higher end of the scale respectively. This implies that the scale is restricted and grouped truncated data typically have a finite number of different categories. For example, we get grouped truncated data, when a

---

[17]Usually treated as interval data with some corrections (Dempster and Rubin, 1983; Heitjan, 1989; Heitjan and Rubin, 1991; Hanisch, 2005)

[18]This subjective nearness of the ordinal and interval scales often leads to the (often inaccurate) approach to treat ordinal data as interval scaled (Mayer, 1971; Knapp, 1990).

person is asked for his or her income with predefined answers like "up to 1000", "2000–3000", "3000–4000", "more than 4000".

As the knowledge of these cut points offers a substantial amount of information, it does not allow us to use every monotonic transformations on the data. This implies that grouped or truncated data contains more information than ordinal data. It is metric data. As this kind of data is generally used to ensure a certain amount of privacy, we will call it *censored* data.

In the subsequent part of the thesis, we will refer to *grouped* data, if we can expect an underlying latent metric variable; but without any information on the cut points.

Ranked data also obtain an order structure and lack the interval structure, but data is completely ordered. Ranked data may be analyzed with the same methods like simple ordinal data, but there are also some specialized techniques (Fürnkranz and Hüllermeier, 2003; Rajaram, Garg, Zhou and Huang, 2003; Shen and Joshi, 2005a; Brinker, Fürnkranz and Hüllermeier, 2006). In recent years, complete orders have been of interest in information retrieval (Shen and Joshi, 2005b). Unfortunately, the terms "ordinal regression" and "ranking learning" are often used synonymously (Shashua and Levin, 2002).

Tutz (1990, 1991) introduced the term *sequential* ordinal data in order to describe ordinal data, where the current class could only be "achieved" by going through other (lower) categories. Examples are: 1) classification of short-term, mid-term and long-term unemployed; 2) current position of a product in the product life cycle in marketing.

The term *hierarchical* data has been introduced by McCullagh and Nelder (1989) in order to describe a scale type, which is similar to sequential ordinal data. The basic idea is, that a population is divided into parts that span a tree with one major branch in a sequence of tests. This model is appropriate either for counting processes and for classifying objects in a hierarchical structure. While in the first case an (underlying) metric scale is implied, in the latter case we have sequential ordinal data.

These are various natural sources of ordinal data. Sometimes, it might be useful to simulate ordinal data. There are well-known techniques for the simulation of categorical and metric data with specified distribution functions (Morgan, 1984). Simulation of non-ranked ordinal data is based on the simulation of nominal data. They only differ during simulation of input-output modelings, because we have to specify a complete cost matrix to simulate an ordinal error function. Potharst and van Wezel (2005) propose a method for ranked ordinal data based on a graph of constraints.

Taxonomy of Ordinal Data

The terms for ordinal data introduced in the former paragraphs are not well defined and we can hardly identify any links between these terms and/or requirements for models analyzing these data. Thus, we are in need of a practicable taxonomy of ordinal data. Unfortunately, there is no complete taxonomy known to the author. Nevertheless, Kampen and Swyngedouw have described a typology of ordinal data which reflects grouped, censored and simple ordinal data (Kampen and Swyngedouw, 2000, p. 99):

Type 1 The *categorized metric variable with known thresholds* corresponds to censored ordinal data.

Type 2 The *categorized metric variable with unknown thresholds* corresponds to grouped ordinal data.

Type 3 The *categorized latent variable (with unknown thresholds)* is based on an unobservable latent variable and categorization is dependent on individual perception. Example: Classification of persons with low, medium or high intelligence.

Type 4 The *semi-standardized discrete variable with ordered categories* is the same as a simple ordinal variable.

Type 5 The *unstandardized discrete variable with ordered categories* is like the simple ordinal variable, but without a inter-rater agreement on the categories of the scale. Examples are agreement questions on questionnaires.

This typology is not sufficient. Let us use the following dimensions to form the taxonomy of ordinal data: Stability of the number of categories, existence of an underlying metric variable, consecutive ordering and possibility of inter-rater agreement (perception of categories). These dimensions do not have any direct mathematical properties, which are useful for a precise definition[19] (cf. section 3.2). Various combinations of these binary criteria are described in Table 2.2.

---

[19]i.e. properties, that would restrict the equivalence class of numerical relatives for ordinal data, cf. subsection 2.1.1.

categories					
number	dependency	underlying	perception	example	terms
fix	none	none	general	hierarchy in hospitals	simple ordinal
fix	none	latent	individual	level of agreement, pain	grouped latent
fix	none	latent	semi-std.	grades	grouped
fix	none	metric	semi-std.	low, mid and high income	grouped
fix	none	metric	general	income classes	censored
fix	sequential	latent	individual	level of education	latent sequential
fix	sequential	latent	semi-std.	developmental status of children	sequential
fix	sequential	metric	semi-std.	phase in product life-cycle	sequential
fix	sequential	metric	general	time of unemployment	sequential
variable	none	latent	semi-std.	search results (relevance)	ranking
variable	none	metric	general	sport results	ranking

Tab. 2.2: Characteristics of ordinal data and examples

Looking at three possible values for scale perception, we can state that "general" allows comparisons between different persons, while "individual" only allows inner-person (maybe in repeated measurements) comparisons. "Semi-standardized" perception allows us to define exact descriptions to classify the objects correctly. Therefore it is possible to ensure a certain amount of comparability (Kampen and Swyngedouw, 2000).[20]

The combinations of variable numbers of categories without an underlying latent metric variable cannot exist, as every total order with an arbitrary number of objects has an underlying metric variable. If such an underlying variable does not exist, we can easily construct it by ranking an infinitely large number of objects. Furthermore, the combination of individual or semi-standardized perception without any underlying variable is not plausible, as there has to be disagreement how to distinguish between the categories.

---

[20]Models for repeated ordinal data are presented by Agresti and Lang (1993); Albert, Hunsberger and Biro (1997); Liu and Agresti (2005); Tutz (2005); Liu and Hedeker (2006).

This implies at least one latent underlying variable as there have to be more realizations of the variable than distinct categories.

The taxonomy shown in Figure 2.2 is suggested to give an overview on different ordinal data. The order of models can be determined as models for data on the higher parts may be used as substitutes for models for data on the lower (more specialized) parts. The root "ordinal data" can be seen as an abstract class, which is instantiated by one of the three types "grouped latent", "simple ordinal" and "ranked latent". The derived types "inherit" the properties from their ancestors. This implies, that models on the lower part of Figure 2.2 are more specialized than the types in the higher part.

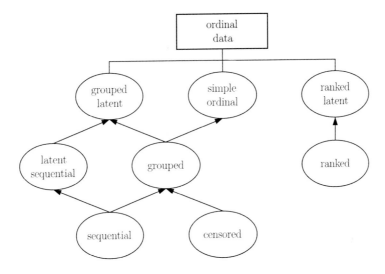

Fig. 2.2: Taxonomy of ordinal data

Measurements of Scale Between Ordinal and Interval

The information gap between ordinal and interval scale is huge. Therefore, we will formulate some additional properties of a data scale that are not necessary to obtain an ordinal scale but restrictive enough for an interval scale.

Let us assume, that the distances between the different categories on an ordinal scale are symmetric to the middle class for uneven numbers of categories and the two middle classes for even numbers of categories respectively. For example, regarding questionnaires for sweets, a possible statement could be: "I like wine gums." with possible answers: "strong disagree, disagree, neutral, agree, strong agree". These formulations are intended to measure data on a symmetric ordinal scale. Unfortunately, no measure of mean or dispersion uses this additional information. Nevertheless, it reduces significantly the number of parameters needed in regression modeling with independent ordinal variables (cf. sections 3.1 and 3.2).

2.2 Machine Learning and Statistics

To classify traditional statistical and machine learning approaches to data analysis, we have to distinguish three aspects: The model, the estimation procedure (target function) and the algorithm that is used to find the optimum of the estimation criterion. Usually, the strength of traditional statistical theory lies in the first two aspects, while the latter has always been ignored as "it can be calculated". In practical settings, many of the traditional "algorithms" are not applicable to large samples and thus the machine learning community developed a lot of different algorithms to solve the same problems; partly with new models and estimation criteria, but also solving well-known estimation problems for old models. Although there are some good comparisons between certain neural network techniques and their statistical counterparts[21], many other areas of machine learning have not been analyzed from a statistical point of view (e.g. clustering), and many statistical models lack a deeper look at the efficiency of their algorithms (especially in nonparametric analysis) or even both (association rules).

Learning and Estimation
The term *estimation* is often called *learning* when talking about neural networks. Essentially, both terms refer to the same concept (Sarle, 1994).[22]

The term *learning* can be specialized with the following pairs of opposites: supervised and unsupervised, online and offline, sequential and batch.

As the terms *supervised* and *unsupervised* are not directly related to estimation, there is no correspondence in classical statistics.[23] These terms

---

[21]i.e. Cheng and Titterington (1994); Ripley (1994a); Warner and Misra (1996); Couvreur and Couvreur (1997); Ripley (1997).

[22]Other "translations" between statistical and neural network terminology can be found in Sarle (1994).

[23]These adjectives describe the model and *not* the estimation or learning process, cf. section 2.3.

indeed describe a structure of the problem by differentiating between problems with known and unknown target (output) of the relationship to be modeled. Thus, the first area of interest is called input-output modeling and the latter structure detection in statistics. Of course, there are different concepts needed to obtain good estimates for the objective functions involved in the problems. The input-output modeling task has been researched in depths within classical statistics and a broad theory is available on that topic. The structure detection task is not that well researched.[24]

The two pairs of opposites *online/offline* and *sequential/batch* are often used as synonyms[25]. However, it would be more exact to distinguish between these two concepts:

Online respective *offline* is used to describe the mode of data obtainment, i.e. data in an offline system is stored completely in the system. In contrast, data "flows" through an online system and cannot be or—for any reason—is not been saved within the system. The usual consequence is that online systems have to process the data immediately and can not retain elapsed data, resulting in high requirements to efficiency of the used algorithms.

The terms *sequential* and *batch* describe the algorithm used for estimation and its use of the data respectively. Sequential algorithms use one observation at a time, while batch algorithms make use of the whole sample at once. Batch algorithms are standard in classical statistics. However, the theory of sequential estimation (Ghosh, Mukhopadhyay and Sen, 1996) and testing explicitly makes use of sequential statistics and algorithms.

Another learning principle often referred to as *batch learning* is a mixture of online/offline and sequential/batch learning. Data batches of a few observations are collected and processed as if it was a pure batch algorithm with the starting values of the preceding optimization step—like in sequential learning. Let m describe the batch size, then for $m = 1$ we obtain an sequential learning algorithm and for $m = n$ with n denoting the sample size we have got batch learning.

Offline data may be learned sequentially or batch, while online data can only be learned sequentially (Sarle, 1997b). The consequence of sequential estimation is a shift from deterministic optimization algorithms to stochastic

---

[24]There are some foundations, for example in clustering (Hartigan, 1975).

[25]For example in Bishop (1995); Saad (1999).

optimization algorithms with loss in efficiency of the calculated estimators (cf. chapters 4 and 5). This also results in different statistical properties. Under light conditions both concepts provide approximately the same estimators, but the convergence is very slow (This is elaborated in chapter 5).

2.3 Statistical Modeling of Input-Output Relations

Statistical modeling will help us to analyze data, but there are many different statistical models, which can be used in data analysis and it is not easy to make the "best" choice. The common idea in statistical modeling is to choose a model according to theoretical considerations and fit it to the data. Therefore, models are collections of plausible "constraints" given for the "real" relationship. Usually, the data will not satisfy these constraints. The differences are interpreted as errors and are represented as random variables.

There are two settings of statistical modeling: Univariate and Multivariate Analysis. Both kinds of models may be parametric, semiparametric or nonparametric. In parametric modeling, every random variable is exactly specified leaving only a few parameters to be estimated. On the other side, in nonparametric models there are no distributions given (as long as there is more than one feasible option) and only structural assumptions like (in-)dependency are made. Within the group of semiparametric models, we can find models that are partly parametric and partly nonparametric (like Projection Pursuit Regression (PPR)). Furthermore, we can build models that are parametric in nature but belong to a large class of models and the remaining parameters have to be estimated as well as one of the models has to be chosen during learning (like in MLP networks, where the number of nodes in the hidden layer is variable).

Within multivariate analysis, we try to establish a connection between the values of two or more variables. This relation can be symmetrical or asymmetrical. Methods like analysis of correspondence or clustering are examples for symmetrical multivariate analysis. In the context of machine learning, symmetric analysis is called unsupervised learning and asymmetric analysis is called supervised learning. Symmetric models are used to find underlying structures or patterns within the data. Examples are clustering, Multi-Dimensional Scaling (MDS) and Principal Components Analysis (PCA).

In asymmetric analysis, the variables are divided into input (or independent) and output (or dependent) variables. The scope of these models is to explain or predict the output variables with the help of the input variables. Depending on the measurement of scale of the output variable(s), we can distinguish between regression for metric scaled output and classification for categorical output variables[26]. Sometimes, models with categorical output variables are called regression models. In this case, a-posteriori probabilities are estimated and no classification takes place. As the a-posteriori probabilities are not nominal, it is called categorical regression instead of classification.

| | | parametrization | | |
		parametric	semiparametric	nonparametric
symmetry	symmetric		structuring	
		finite mixture[27]	*ART* *k-means*	*nearest neighbor*
		LDA		*SVM, CT*
	asymmetric	explanation		prognosis
		GLM	*MLP* *PPR*	*RT*

Tab. 2.3: Morphologic matrix: Multivariate analysis

In Table 2.3 popular statistical models (grey) are arranged according to the two dimensions "parametrization" and "symmetry" besides their purposes "structuring", "explanation" and "prognosis" (black). Parametric models are better suited for explanation, while nonparametric model have got a lower prediction error in general(Breiman, 2001). Although parametric models often suffer from making worse predictions, they are much easier to be interpreted and their results are more likely to be predicted.[28] This may be one of the reasons, why they are very popular both for traditional theory verification and for forecasting real world phenomena (cf. e.g. Wang (1995)). Both tasks

---

[26]The term classification may be used for symmetric *and* asymmetric analysis as the theoretical model in the background is the same, see subsection 2.3.2.

[27]Finite mixture models are used as the basic model for theoretical classification. If it is identifiable, which implies a complete specification of the involved distributions, it is a parametric estimator.

[28]Counter-examples are Classification and Regression Tree (CART) which provide explanations for their predictions.

distinguish between input and output variables (especially in prediction), as one (or more) variable(s) should be explained/predicted with the help of the other variables and that is why the variables have to be treated asymmetrically (see Table 2.3). Both models—parametric and nonparametric—are useful for detection of underlying structures or patterns, but nonparametric models are usually used because we do not know much about the variables and their distributions in particular.

The methods for asymmetric analysis in Table 2.3, called Linear Discriminant Analysis (LDA) and GLMs, are closely related to certain aspects of MLPs (cf. subsections 2.3.2 and 2.3.3). PPRs are the statistical counterpart of MLP networks, as they are semiparametric models and they describe a function which consists of additively combined simple functions (Friedman and Stuetzle, 1981). However, the semiparametric characteristic of PPR results from nonparametric simple functions rather than from a parametric combination of fixed transfer functions (Ripley, 1993; Cheng and Titterington, 1994). Support Vector Machine (SVM) and CART are briefly described in subsection 3.2.3.

2.3.1 Regression

In general, the term regression is used to describe the conditional mean of a random variable $Y \in \mathbb{R}$ depending on a random vector $\boldsymbol{x} \in \mathbb{R}^d$ by a function f:

$$\mathbb{E}(Y|\boldsymbol{x}) = f(\boldsymbol{x}) \tag{2.3}$$

with $f : \mathbb{R}^d \to \mathbb{R}$.

Now, choosing an approach with additive error $\epsilon \in \mathbb{R}$ with $\mathbb{E}(\epsilon) = 0$ for a metric Y

$$Y = f(\boldsymbol{x}) + \epsilon$$

we obtain the following properties:

i) $\mathrm{Cov}\,(f(\boldsymbol{x}), \epsilon) = 0$

ii) $\mathbb{E}(f(\boldsymbol{x})) = \mathbb{E}(Y)$

iii) $\mathbb{R} \ni R^2 = 1 - \frac{\mathrm{Var}(E)}{\mathrm{Var}(Y)}$

This basic principle of regression is not useful for applied statistics as long as we do not have a really huge number of i.i.d. samples, because we have to estimate the distribution $P(Y|\boldsymbol{x})$ to determine $f(\boldsymbol{x})$. That is why, we have to restrict the class of regression functions $f(\boldsymbol{x})$ by using a well-chosen model.

Models

The function f is often assumed to be linear in a tuple of parameters $\boldsymbol{\beta} = (\beta_0, \dots, \beta_d)$ building a function of the type

$$f(\boldsymbol{x}) = f(\boldsymbol{x}|\boldsymbol{\beta}) = \beta_0 + \beta_1 x_1 + \cdots + \beta_d x_d$$

This model is very restrictive as it is restricted to linear relationships. Of course, the function f does not have to to be parametric or linear[29], but in this thesis, f is assumed to be parametric with

$$f(\boldsymbol{x}) = f(\boldsymbol{x}|\boldsymbol{w}) \qquad (2.4)$$

where $\boldsymbol{w}$ describes the parameters, which are often called "weights" in neural network models, and f is nonlinear.[30]

Regression is a problem involving a lot of facets and thus it is covered by various books in more detail.[31]

Estimation

Equation 2.4 from an i.i.d. sample $(\boldsymbol{x}_i, y_i)_{i \in \mathbb{N}}$ with $\boldsymbol{x}_i \in \mathbb{R}^d$, $y_i \in \mathbb{R}$ is written as

$$y_i = f(\boldsymbol{x}_i|\boldsymbol{w}) + \epsilon_i$$

where y_i denotes the i-th output value, $\boldsymbol{x}_i$ the corresponding input vector and $\epsilon_i \in \mathbb{R}$ an appropriate error term with $\mathbb{E}(\epsilon) = 0$.

The method of Least Squares (LS) is standard for the estimation of $\boldsymbol{w}$ (Sen and Srivastava, 1990). Compared to maximum likelihood its advantage is the "model-free" approach concerning the error term.

---

[29] e.g. regression trees (Breiman, Friedman, Olshen and Stone, 1984).

[30] The non-linearity has to be restricted too, i.e. we will use a special kind of non-linearity which is described later (see chapter 4).

[31] e.g. Sen and Srivastava (1990), McCullagh and Nelder (1989), Green and Silverman (1994) and Seber and Wild (2003) describing parametric models and Hastie, Tibshirani and Friedman (2001), Breiman, Friedman, Olshen and Stone (1984) and Cristianini and Shawe-Taylor (2000) for nonparametric models.

In general, the LS estimation starts with the MSE function

$$LS(\boldsymbol{X}, \boldsymbol{y}|\boldsymbol{w}) = \sum_{i=1}^{n} (y_i - f(\boldsymbol{x}_i|\boldsymbol{w}))^2 \overset{!}{=} \min_{\boldsymbol{w}} \,, \qquad (2.5)$$

which can be solved for differentiable $f(\cdot|\boldsymbol{w})$ by

$$\begin{aligned}
\frac{d}{d\boldsymbol{w}} LS(\boldsymbol{X}, \boldsymbol{y}|\boldsymbol{w}) &= \frac{d}{d\boldsymbol{w}} \left[\sum_{i=1}^{n} (y_i - f(\boldsymbol{x}_i|\boldsymbol{w}))^2 \right] \\
&= \frac{d}{d\boldsymbol{w}} \left[\sum_{i=1}^{n} \left(y_i^2 - 2y_i f(\boldsymbol{x}_i|\boldsymbol{w}) + (f(\boldsymbol{x}_i|\boldsymbol{w}))^2 \right) \right] \qquad (2.6) \\
&= 2 \sum_{i=1}^{n} (f(\boldsymbol{x}_i|\boldsymbol{w}) - y_i) \frac{d}{d\boldsymbol{w}} [f(\boldsymbol{x}_i|\boldsymbol{w})] \,.
\end{aligned}$$

Choosing the quadratic error measure is advantageous, because it is differentiable and under the additional assumption of Gaussian error, it renders the ML-estimator. Without any further assumptions on the error distribution, the LS-estimator can be shown to be BLUE (Sen and Srivastava, 1990). This result is obtained under the assumption of controlled explanatory variables. In this case, we allow the explanatory variables to be generated by a random process, we have to assume additionally that the errors $\boldsymbol{\epsilon}$ are stochastically independent (Judge, Griffiths, Carter Hill, Lütkepohl and Lee, 1985). This should be assumed for the rest of this thesis[32].

2.3.2 Classification

The classification problem involves a response variable Y taking values in a finite set C of disjoint categories[33] depending on a tuple of explanatory variables $\boldsymbol{x} \in \mathbb{R}^d$.

In general, classification problems have the advantage that the error-model for misclassification is obviously a Dirac measure[34]. This binary error measure

---

[32]Extension of the results to stochastically dependent variables can be obtained analogously to general nonlinear regression (Seber and Wild, 2003, chap 6).

[33]There are also classification methods called fuzzy which explicitly allow non-disjunct categories, see, for example, Halgamuge and Wang (2005) for some recent developments.

[34]$\delta_x(A) = 1$ iff $x \in A$ and zero otherwise.

allows us to formulate an theoretically optimal model for classification. In contrast to regression, the minimal classification error for a completely specified problem is given by a fix point which is called Bayesian classification.

Core of the Bayesian classification is the Bayesian decision rule

$$d_B : \mathbb{R}^d \to C$$
$$d_B(\boldsymbol{x}) = \operatorname*{argmax}_c P(c|\boldsymbol{x}) \tag{2.7}$$

based on the a-posteriori probabilities $P(c|\boldsymbol{x})$ for categories $c \in C$. Using Bayes formula we can calculate the a-posteriori probabilities via

$$P(c|\boldsymbol{x}) = \frac{p(\boldsymbol{x}|c)P(c)}{p(\boldsymbol{x})} \quad . \tag{2.8}$$

Therefore, we have to know the exact distributions of the feature vector for each class—the a-priori probability that an unknown object may belong to class c—and the probability of a specific feature Vector $p(\boldsymbol{x})$. Unfortunately, these quantities are usually unknown and have to be estimated.

It can be shown for the simple error function that the Bayes decision rule minimizes the risk for choosing class c if the object belongs to class α is

$$R(c|\boldsymbol{x}) = 1 - P(\alpha|\boldsymbol{x}) \quad . \tag{2.9}$$

Therefore, maximizing the a-posteriori probability $P(c|\boldsymbol{x})$ is associated with minimizing the risk of misclassification (Ripley, 1996; Duda, Hart and Stork, 2000).

There also exists a maximum likelihood decision rule:

$$d_{ML}(\boldsymbol{x}) = \operatorname*{argmax}_c p(\boldsymbol{x}|c)$$

Assuming equal a-priori distributions—which might be feasible under the complete uncertainty assumption—the Maximum Likelihood (ML) and the Bayes decision rules are equal.

Besides the common binary error of misclassification, we can introduce misclassification costs $R_{i,j}$ denoting the costs that result from choosing class i instead of class j. Then decision rule (2.7) changes to

$$d(\boldsymbol{x}) = \operatorname*{argmin}_{c} R_{\ell,c} \sum_{\ell \in C} P(c|\boldsymbol{x}) \ . \tag{2.10}$$

Pure classification models or algorithms like LDA or SVM are in need of certain adjustments to incorporate misclassification costs.[35] This is not the case for categorical regression algorithms, as they only approximate the a-posteriori probability and thus the rules d_B and d_{ML} may be used with or without misclassification costs.

Unfortunately, the Bayes decision rule—although optimal—has two major drawbacks in practical application. First, the used distributions have to be known. If they are unknown, they have to be estimated. In applications, the data sets are usually not large enough to provide an estimator for a nonparametric estimation. Secondly, this rule is only applicable for known error values. While in nominal classification we may argue that all errors have equal costs, this would be impossible for ordinal categories. This implies that there are two cases of ordinal classification: with and without cost information (cf. section 3.2).

Now there are two major approaches to classification: First, the direct way via estimation of the decision boundary and secondly, the more indirect way by estimating the a-posteriori probabilities. The first approach is usually called classification and the latter one categorical regression. Estimation of the decision boundaries can be done with the help of methods like LDA and (more complex) SVM which are based on the highly intuitive large margin principle. These methods are nonparametric and do not explicitly estimate the a-posteriori probabilities and use another (more or less) linear decision rule (Duda, Hart and Stork, 2000). Another approach is represented in its simple form by logistic regression, which estimates the a-posteriori distributions. A generalization to multiple categories is done via GLMs which are introduced in the following subsection.

There are different attempts to find a superior classification approach for different settings by comparing LDA and logistic regression. Obviously, LDA

---

[35]Examples can be found in Grouven, Bergel and Schulz (1996) and Schiffers (1997) among others.

performs better for multivariate normal class distributions (Efron, 1975), but it is not very robust to deviations from this assumption. This is why, especially for non-metric data and other obviously non-metric settings, the better performing linear discriminant is suggested (Press and Wilson, 1978).

2.3.3 Generalized Linear Models

Nelder and Wedderburn (1972) have introduced Generalized Linear Models (GLMs) and McCullagh and Nelder (1989) describe them in detail. The basic idea of generalized linear models is to provide a common framework for linear regression analysis with various distributions of the output variables within the exponential family. Special cases are the classical regression model with Gaussian error, logistic regression for binary classification and log-linear models whereas the latter are used for count data.

Thus, GLMs are appropriate for regression *and* classification tasks and are (cf. chapter 4) closely related to MLP networks. This is why they should be described in more detail than other regression or classification models.

Model

Let us assume a parametric model for the distribution of Y with expectation μ dependent on regressors x, i.e. we would like to model the constraint expectation $\mathbb{E}(Y|x)$. Let us further assume that there is a systematic component η and we can construct a linear relationship between the explanatory variables X on the one hand and η on the other:

$$\eta = \beta_0 + \langle x, \beta^- \rangle \tag{2.11}$$

where $\beta^- \in \mathbb{R}^d$ only contains the weights for the different regressors, but not the constant parameter β_0. Further, let us assume that there is a (distribution dependent) link function between μ and η. As η has to be unconstrained, choosing a link function that represents the natural parameters of one of the distributions from the exponential family is an obvious choice.

Commonly used link functions are the identity for univariate regression, the logit function for binary classification and the logarithm for Poisson regression (McCullagh and Nelder, 1989) (see Table 2.4).

name	distribution(Y)	canonical link		model ($\mathbb{E}(Y)$)
standard regression	Gaussian	identity	μ	$\beta_0 + \langle x, \beta^- \rangle$
logistic regression	Binomial	logit	$\log\left[\dfrac{\mu}{1-\mu}\right]$	$\dfrac{\exp[(1, x^t)\beta]}{1 + \exp[(1, x^t)\beta]}$
Poisson regression	Poisson	log	$\log \mu$	$\exp[(1, x^t)\beta]$
logistic regression	Multinomial	logit	$\log\left[\dfrac{\mu_c}{1 - \sum_{r=1}^{k-1} \mu_r}\right]$	$\dfrac{\exp[(1, x^t)\beta_c]}{1 + \sum_{r=1}^{k-1} \exp[(1, x^t)\beta_r]}$

Tab. 2.4: Examples for GLMs

It can be shown for these link functions that the statistic $X^t Y$, with components

$$\sum_i x_{ij} Y_i \quad j = 1 \ldots d, \tag{2.12}$$

is sufficient for estimating the natural parameter vector θ of the random component of Y (McCullagh and Nelder, 1989).

A direct generalization of the logistic regression is the multi-categorical logit which is build from a binary logit model by replacing the expectation μ by a vector μ forming the model

$$\mathbb{E}(Y_r | x) = \frac{\exp[\beta_{0r} + \langle \beta^-, x \rangle]}{1 + \sum_{r=1} k - 1 \exp[\beta_{0r} + \langle \beta^-, x \rangle]} \quad r = 1, \ldots, k-1 \tag{2.13}$$

In the case of polytomous categorical regression we can further generalize the canonical link function to every monotone increasing function with asymptotes zero and one for infinite large negative or positive numbers respectively (i.e. a Cumulative Distribution Function (CDF)). This opens up the model to link functions like the probit, the complementary log-log or the very flexible power family (McCullagh and Nelder, 1989). Although the link functions do not influence the basic (linear) form of the decision boundaries, they have a certain influence on their positions as they differently weight the errors. This may be appropriate, if the distribution of an underlying continuous variable with a threshold for a binary outcome is known (Agresti, 2002). Particularly, the link function is neither unique nor even necessary for nonlinear regression on a multinomial variable (Cox, 1984). This will be discussed in section 4.5 and play a major role in section 7.4.

Estimation

Using GLMs for linear classification results into a completely specified parametric model and estimation via maximum likelihood is possible. Thus, for estimation maximizing the log-likelihood function

$$\log L(\boldsymbol{y}|\boldsymbol{X}, \boldsymbol{\theta}) = \sum_i \log f_i(y_i|\boldsymbol{x}_i, \boldsymbol{\theta}) \qquad (2.14)$$

with $f(\cdot)$ denoting the chosen distribution function in $\boldsymbol{\theta}$ will provide the ML-estimator. $\boldsymbol{\theta}$ may consist of the expectation parameter $\boldsymbol{\eta}$ and some further side-parameters for distortion, skewness and the like.

As the minimum of (2.14) often has to be calculated by a numerical optimization procedure, the most popular of those, the Iteratively Re-weighted Least Squares (IRLS) algorithm, is introduced (McCullagh and Nelder, 1989, p. 40f.). It is based on the work by Nelder and Wedderburn (1972)and it is closely related to the Fisher-Scoring algorithm (Rao, 1973). It has first been implemented for the software Generalized Linear Interactive Modeling (GLIM) (Baker and Nelder, 1978). The algorithm works iteratively on the estimates $\hat{\boldsymbol{\theta}}$ to find an at least local minimum of the likelihood function. Let $\hat{\boldsymbol{\eta}}_0$ be the current estimate of the predictor function $\hat{\boldsymbol{\eta}}_0 = f(\boldsymbol{x}|\hat{\boldsymbol{\theta}}_0)$ for all samples. Let further $\hat{\boldsymbol{\mu}}_0$ be the corresponding value derived from the link function $\boldsymbol{\eta} = g(\boldsymbol{\mu})$. Calculate the adjusted dependent variate with value

$$z_0 = \hat{\boldsymbol{\eta}}_0 + (\boldsymbol{y} - \hat{\boldsymbol{\mu}}_0)\left(\frac{d\boldsymbol{\eta}}{d\boldsymbol{\mu}}\right)_0$$

where the derivative of the link is evaluated as $\hat{\boldsymbol{\mu}}_0$. $\boldsymbol{z}$ is a linearized form of the link function applied to the data. The quadratic weight is defined by

$$\boldsymbol{W}_0^{-1} = \left(\frac{d\boldsymbol{\eta}}{d\boldsymbol{\mu}}\Big|_0^2 \boldsymbol{V}_0\right)$$

where $\boldsymbol{V}_0$ is the variance function of the parameter evaluated at $\hat{\boldsymbol{\mu}}_0$. Now, we have to regress $\boldsymbol{z}_0$ on the predictors $\boldsymbol{X}$ with weight $\boldsymbol{W}_0$ to give new estimates $\hat{\boldsymbol{\theta}}_1$ and from these the new estimate $\hat{\boldsymbol{\eta}}_1$ of the predictor.

In this formulation, the concrete algorithm for the regression is left open and we would like to refer to McCullagh and Nelder (1989, p. 81ff.) for some discussions on that topic for linear regression functions. As we are in need of algorithms for nonlinear regression in the later chapters, this topic is not discussed here.

Goodness-of-Fit Testing

There exist two major statistics for measuring the goodness-of-fit for a GLM.[36] First, the deviance (likelihood-ratio) statistic

$$D(\boldsymbol{y}, \hat{\boldsymbol{\mu}}) = -2 \sum_{i=1}^{I} \left(l_i(\hat{\boldsymbol{\mu}}_i) - l_i(\boldsymbol{y}_i) \right) , \tag{2.15}$$

where $l.$ denotes the likelihood for the ith covariates pattern and n_i the number of observations with covariates pattern i. Secondly, the generalized Pearson X^2-statistic

$$X^2 = \sum_{i=1}^{I} n_i \sum_{r=1}^{k} \frac{(y_{i,r} - \hat{\mu}_{i,r})^2}{\hat{\mu}_{i,r}} . \tag{2.16}$$

For $\frac{n_i}{n} \to \lambda_i \in (0,1]$, both statistics are approximatively χ^2 distributed with $(I-1)(r-1) - d$ degrees of freedom (McCullagh and Nelder, 1989, p. 34).

These statistics are nearly unusable for $I \to N$ which is often the case for continuous explanatory variables, as the approximation of the χ^2 distribution does not hold.

2.4 Validation and Model Selection Problems

It is well known that there is no general best model for regression and/or classification problems (Duda, Hart and Stork, 2000). Therefore, we may have to try out many classifiers and choose the "best" classifier for our actual problem. There are two major concepts interrelating: prediction error and model complexity.

A more complex model can better be adjusted to the data (see Figure 2.3), but it is often over-fitted and has no good generalization capability. Unfortunately, the model complexity cannot be chosen on a continuum and nobody really knows, if there is a global minimum: It just describes a tendency, that has been found in many applications.

---

[36] As for the remainder of this thesis the multinomial classification problem will be of interest, the statistics are constructed for this case. Further information on GLM and appropriate statistics can be obtained from McCullagh and Nelder (1989, p. 33ff.).

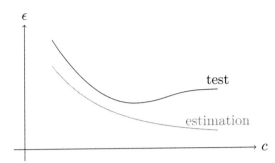

Source: Hastie, Tibshirani and Friedman (2001, p. 194)

Fig. 2.3: Test and estimation sample error in relation to model complexity

The term generalization capability is used to describe the performance of a model and/or its estimation procedure with unknown data. This generalization capability is measured with the help of test data to estimate the prediction error. Therefore, we have to divide the data set into two parts, the first part for learning and second part for testing. Furthermore, as we may want to choose the best available model, we are also in need of some validation data. Thus, all in all three data sets for learning, model selection and model assessment are required.

We have to try out various models with different complexity for choosing the right level of complexity.[37] The easiest way would be, if we had a complete order on the models regarding their complexity. Then we could "add" complexity until the prediction error starts rising. Unfortunately, there does not exist such an order for every thinkable statistical model because no complete measurement of complexity is known. Only if we stay within a model type we can control complexity in this way. For example, we can control the complexity[38] with the number of nodes in the hidden layers of MLP networks (cf. section 4.1).

This technique for model selection leads to the data analysis process shown in Figure 2.4. We have to mention especially that this recursive structure is not applicable to online learning, as the training data has to be "reused" in all iterations and therefore has to be accessible all the time. This problem is

---

[37]We will use the "iterative selection of a model". A second possibility to deal with model complexity is "penalizing the fit" (Ripley, 1996, p. 60).

[38]Again, only discrete and not continuously.

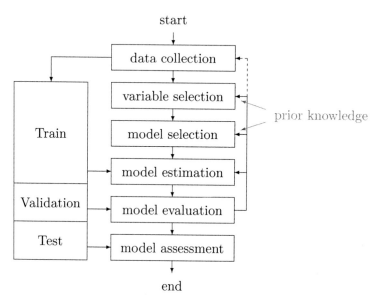

Source: Duda, Hart and Stork (2000, p. 14) and Hastie, Tibshirani and
Friedman (2001, p. 196)

Fig. 2.4: Data analysis process

one of the major arguments for the use of sufficient statistics in online learning
tasks (cf. chapter 6).

In general settings, there are different techniques for estimating the prediction
error. Namely techniques like cross-validation or bootstrapping are used (Hastie,
Tibshirani and Friedman, 2001). For MLP networks, cross-validation is more
popular (Bishop, 1995; Ripley, 1996) but it is problematic (Fine, 1999).

The basic idea of cross-validation is to divide the sample into V pieces, using
$V - 1$ of these pieces for estimation and one piece for assessing the performance
of the resulting model. The resulting V different configurations can be used to
determine the test error of the model. Cross-validation is very useful for small
sample sets, when separation into a test and independent estimation sample
is problematic. This method is computationally expensive and not useful for
large samples (Ripley, 1996, p. 69).

It is impossible to determine the best classifier model a-priori (Duin, 1996).
As it is even impossible to estimate the right model complexity, we are in need
of a method for comparing the complexities of various possible classifiers. The

concept of the Vapnik-Chernovenkis (VC) dimension provides a theoretical upper bound for the necessary model complexity (Vapnik, 1998). Unfortunately, the VC dimension is difficult to be calculated (Hastie, Tibshirani and Friedman, 2001, p. 212), especially for MLP networks (Fine, 1999, p. 280f.). Although the concept of the VC dimension reduces the space of possible models significantly in theory, it is of low practical use because the excluded models are usually ignored anyway.

2.5 Neural Networks

Some authors use a biologically based motivation for their artificial neural network books (Zell, 1994; Rojas, 1996; Haykin, 1999). They described biologic— i.e. natural—neural networks as nonlinear, parallel processing, online learning, networked systems which they try to simulate with the help of artificial neural networks. Others do not even try to define the term "neural networks", but start directly with a constructive description of the analyzed network type (Hecht-Nielsen, 1990; Bishop, 1995; Duda, Hart and Stork, 2000).

Haykin (1999, p. 2) offers a definition adapted from Aleksander and Morton (1990), also used by Ripley (1996, p. 4):

> *A neural network is a massively parallel distributed processor that has a natural propensity for storing experiential knowledge and making it available for use. It resembles the brain in two aspects:*
>
> *1. Knowledge is acquired by the network through a learning process.*
>
> *2. Inter-neuron connection strengths known as synaptic weights are used to store the knowledge.*

This definition is neither complete (e.g. Kohonen networks are excluded) nor very precise. In addition Ripley (1996, p. 4) emphasizes the online learning algorithm and the semiparametric model of neural networks. Nevertheless, this definition is sufficient. Statistically speaking, it introduces the following characteristics of MLP networks: Network-like structure of the graphical representation of the model, nonlinearity, parallel calculation, semiparametric model and sequential estimation (online learning).

Using these basic ideas in standard statistical models—as proposed for the graphical network structure by Ripley (1994a)—result (especially in looking at the online learning principle) in new techniques useful in a data mining context. Thus, these extended statistical models con be used for large amounts of data and in online learning tasks.

Neural networks are often used in a data mining context. Tan *et al.* mention three characteristic properties for techniques belonging to and being appropriate for data mining: 1) scalability, 2) high dimensionality of the data, 3) heterogeneous and complex data. Further they refer to 4) complex data ownership (privacy) and distribution over many different sources, and 5) non-traditional analysis (Tan, Steinbach and Kumar, 2005, p. 4f.). Characteristic 5) is a result of the other challenges as most of the traditional methods do not scale well.

Characteristics 1) to 3) indicate that neural networks are useful in a data mining task, because they are especially designed for data having these characteristics. Using the parallel computation ability of neural networks, they are scalable and therefore well suited for high dimensional data. From a more statistical point of view, the relative small number of parameters obtained by using layers, the curse of dimensionality is handled very efficiently.

The forth characteristic is gaining more and more attention, because information is becoming more worthy. This leads to data that is "censored" in many ways, often by reducing metric variables to ordinal ones, which founds the challenge of extracting as much information left within the data as possible without over-interpreting them. Thus, appropriate methods are required to deal with ordinal data in the context of neural networks. Unfortunately, the term "neural network" is not well-defined covering various different approaches. These are only loosely coupled by the idea of simulating natural connectional structured.

2.5.1 Characteristics of Neural Networks

Graphical Representation of the Model

Neural networks are often represented as graphs in which connections represent weights and nodes contain a (nonlinear) transformation $\mathbb{R}^n \to \mathbb{R}$ with n representing the number of incoming connections. This representation can be used to distribute the calculation tasks in order to use the potential parallel computation (Haykin, 1999).

Of course, this kind of representation for a statistical model is not specific to neural networks. For example, Anders (1997) presents graphical representations for linear regression, logistic regression and GLM.

The following symbols[39] are used throughout the thesis for representing statistical models in order to show the associated parallelization potential:[40]

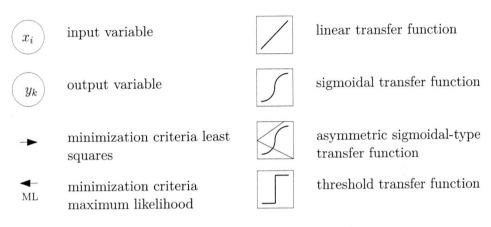

Fig. 2.5: Symbols used for graphical representations of neural networks

Nonlinear Semiparametric Models
Neural networks (especially MLP networks) are combinations of linear and nonlinear functions and therefore nonlinear in total. In the case of feed-forward networks, the replacement of all nonlinear functions by linear ones would lead to ordinary linear regression. The nonlinearity is usually (i.e. in MLP networks) restricted by only using smooth and often even only totally differentiable functions.

The term semiparametric model is used for models that consist of parametric parts and nonparametric parts. Often the error term is the only nonparametric part of a semiparametric model (i.e. general linear regression with unbiased and homoscedastic errors in contrast to classical linear regression with Gaussian error). Besides the error term, the functional relationship may also be semiparametric like the PPR by Friedman and Stuetzle (1981) which consists of a

---

[39]Symbols are inspired by (Sarle, 1994), but different in detail.
[40]Sigmoidal functions are functions with a s-shaped graph. They are introduced in detail in subsection 4.1.1.

parametric "mixture" of nonparametric functions. Neural networks are often classified as semiparametric (Ripley, 1996; Hastie, Tibshirani and Friedman, 2001), but this is not always correct. For example, a simple MLP network with gaussian error is parametric because the model (and thus the function f) is parametric.[41] A MLP network with an adaptable model (through adding and/or removing nodes) is semiparametric whether we assume a special error distribution or not.

Sequential Estimation of Parameters

One of the most interesting aspects of neural networks is the online learning property which is often seen as *the* reason for neural networks to belong to the area of artificial intelligence (Ripley, 1996; Haykin, 1999). Like "natural" intelligence neural networks can learn when they are used for prediction. This allows a kind of dynamic adaption to changing environments, i.e. adaption of the estimated relationship.

One of many possibilities for finding the parameters, which minimize the error function, is gradient descent (Bishop, 1995, chap. 7). Its updating formula for a parameter w with pattern $n + 1$ and its error function E^{n+1} and the learning rate η is:[42]

$$w^{(n+1)} = w^{(n)} - \eta \frac{\partial E^{(n+1)}}{\partial w} \qquad (2.17)$$

Parallel Computation

One of the most important contributions of neural networks to data analysis is the consequently parallel design, which is represented in the graphical network structure. Neural networks have been optimized in order to achieve a high benefit from calculation parallelization (Rumelhart and McClelland, 1986).

Quinn (2003) distinguishes between data parallelism, functional parallelism and pipelining:

- Data parallelism is useful, if there are a lot of instances that have to be processed by the same operations, i.e. in the context of statistics a method allows data parallelism, if the samples may be divided into disjoint sub-samples, results are calculated for these sub-samples independently and

---

[41]This assumption is not unusual (White, 1994), particularly it is needed in testing (White, 1989a).

[42]This rule is generalized to multiple parameters in section 4.2.

these results are merged afterwards. This division of the sample may be done pattern-wise or attribute-wise, whereas the latter has got a lower scalability potential because the number of attributes is fix.

- Functional parallelism is possible, if there are different tasks to be done for different parts of the data, i.e. estimation of the mean and the dispersion of a Gaussian distribution.

- Pipelining is an alternative to data parallelism for more complex operations on each item without dependence on other items. Sequential operations to be performed for one instance are distributed and the object will "travel" between the distributed operations to be done. This can be compared to line production in the automobile industry.

There are four different approaches to parallel computation in feed-forward neural networks (e.g. MLP networks) (Tørresen and Tomita, 1998):[43]

- Training session parallelism (none)
  The complete pattern set is learned by different neural networks (initial weights and/or models). This can speed up the time needed for designing an appropriate neural network model and/or finding the global minimum of the error function. But this kind of parallelization does not scale in terms of pattern set size or network model size.

- Training set parallelism (data parallelism of the sample)
  Splitting the training set across the processors with each processor holding a local copy of the complete model (weights). It is not useful for online learning because the weights have to be updated too often forming a bottleneck at the central weight holding and distributing unit. But if it is used in batch learning, it is the most efficient parallelization technique, especially when computer networks (clusters) are used.

- Pipelining (pipelining)
  Pipelined calculation in neural networks is done by dividing them layer-wise on different processors. It is only useful for batch learning because the weights are updated *after* the next pattern has passed the first layer and therefore the update would be applied too late.

---

[43]The related concept by Quinn is noted in parentheses.

- Node parallelism (functional and data parallelism of the attributes)
  All nodes (or connections) in one layer are calculated on their own
  processors. This is the most flexible (and most parallel) architecture and
  useful for online learning. Nevertheless, because of its huge amount of
  communication in relation to the computation effort it is only efficient,
  if common memory is available for all processing units. Furthermore,
  it does not scale well as there is an upper bound given by the implicit
  parallelism of the neural network, i.e. the amount of parallelization grows
  with the size of the network but it cannot grow with the number of
  patterns.

Looking at the graphical representation of a neural network model, the node
parallelism seems to be the most intuitive approach to parallel computation in
neural networks but requires a fast communication for being efficient. Training
set parallelism can be realized with much less communication cost.

Neural networks are usually optimized for pipelining or node parallelism
for their simulation on Multiple Instruction Single Data (MISD) Processors.
Training set parallelism is the only concept of parallel computing which is not
possible for many neural network models because of their stochastic estimation
principle. The main reason for this property is the order dependency of the
stochastic estimation procedure from the pattern set.

2.5.2 Using Basic Neural Network Concepts in Linear Regression

Linear regression can be looked at as a neural network with one single, linear
neuron (cf. Figure 2.6).

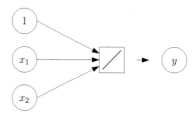

Source: Anders (1997)

Fig. 2.6: Linear regression using the graphical representation of neural networks

The sequential backpropagation algorithm may be used to update the parameters in order to obtain the dependence of the result on the ordering of the sample. Online learning are also (and statistically "better") realized via defining a sufficient statistic with a dimension that is independent from the sample size n. But using these statistics is *not* an idea of neural networks and therefore discussed in section 6.1

Using equation

$$\delta_k = g_k'(a_k^{(n)})\frac{\partial E^{(n)}}{\partial y_k}$$

for updating the parameters and the quadratic error function leads to the following equations:

$$\delta = g'(a)\frac{\partial E}{\partial y}$$

$$= 2\left(y - \left(\beta_0 + \sum_j \beta_j x_j\right)\right) \tag{2.18}$$

Thus, using a simple "gradient descent" the updating rules are (cf. equation (2.17)) in the case of sequential estimation:

$$\beta_j^{(n+1)} = \beta_j^{(n)} - \eta \cdot 2\left(y^{(n+1)} - \left(\beta_0^{(n)} + \sum_j \beta_j^{(n)} x_j^{(n+1)}\right)\right) \cdot x_j^{(n+1)} \tag{2.19}$$

And the batch estimation algorithm would be:

$$\beta_j^{(m+1)} = \beta_j^{(m)} - \eta \cdot 2\sum_i \left(\left(y^{(i)} - \left(\beta_0^{(m)} + \sum_j \beta_j^{(m)} x_j^{(i)}\right)\right) \cdot x_j^{(i)}\right) \tag{2.20}$$

It is easy to see that the estimator is biased, as the starting value of the algorithm will not be equal to the value of the estimator. This and the different variances of the sequential and the batch estimator are analyzed in depth in chapter 5.

In the case of linear regression, no numerical algorithm is necessary because there is an algebraic closed form of the solution to the normal equations. Therefore, a sufficient statistic is easily defined and updated sequentially. This

example introduces the basic idea of using sequential sufficient statistics that
is worked out in chapters 6 and 7 in particular.

Adaptive Learning

The learning rate η has two[44] functions in controlling the "learning" process:

a) Technically, it is a control parameter for the numerical optimization algorithm used for estimation, and

b) semantically, it controls the adoption speed of a regression function for an
   unstable (time-dependent) relation.

This double role of the parameter is not desirable. One of the aims of this
thesis is to split up η into its technical and its semantical part. The weighting
of this "moving average" analysis is dependent on the differences between the
estimated and the observed value of y. Thus, after dividing it into its technical
and semantical parts, the specific adaption is not easy to be simulated by the
two new parameters.

---

[44]These two functions are given for online learning, only the technical interpretation holds
in offline learning.

3 Ordinal Data in Input-Output Models

Using the additional information of ordinal data may improve the quality of the generated classifiers. Although this sounds well in theory, we should have a look at some studies, which have compared ordinal models with their corresponding nominal counterparts.

Campbell and Donner have proven that the ordinal model by Anderson (1984); Greenland (1985) is superior in terms of the asymptotic relative classification efficiency to the standard multinomial logistic model (Campbell and Donner, 1989). The relative efficiency ranged between 0.073 and 0.542 and thus we could state that in this case, an ordinal model has only half the classification errors than the corresponding nominal model. As this result holds for simple misclassification error, they did not take a look at it with (possible) greater costs for misclassification in categories at different ends of the scale.

In a simulation study, Campbell et al. have shown that ordinal models[45] do not have any advantage to ordinary classification procedures (linear discriminant, multinomial logistic). However, they have simulated data which did not match the assumptions for ordinal models. This mistake has been corrected in the subsequent paper by Rudolfer, Watson and Lesaffre (1995). They obtained slightly better results in term of misclassification error rates for ordinal models, if the data matches the assumptions of the ordinal model.

These more theoretical results have also been confirmed in real data analysis (Hastie, Botha and Schnitzler, 1989; Taylor and Becker, 1998). There is also a study comparing logistic regression, LDA, ordinal logistic regression and ordinary linear regression models for ordinal data collected via a health assessment questionnaire. Surprisingly, the linear regression performed second best, which would imply that (in this particular case) linear models (although

---

[45]ordinal logistic (Anderson, 1984), continuation-ratio, proportional odds (Walker and Duncan, 1967; McCullagh, 1980), (Anderson and Philips, 1981)

the popular assumption of equally distances between categories is very heroic) may be more useful for this kind of data than nominal models are. The ordinal logistic regression model provides the lowest misclassification error and thus would be preferable to linear regression (Norris, Ghali, Saunders, Brant, Galbraith, Faris and Knudtson, 2006).

Agresti (1986) describes the different facets of ordinal data analysis and more recently, these are discussed in combination with GLM in Agresti (1999).

The taxonomy of ordinal data (see Figure 2.2) does have a strong influence on data analysis (Kampen and Swyngedouw, 2000). The different approaches to ordinal data modeling only partly represent alternative approaches for the same kind of ordinal data.

Unknown Misclassification Costs

Referring to the situation of the Bayes classification (cf. subsection 2.3.2) we can think of ordinal data as categorical data with unknown misclassification costs whereas these costs are ordered. That is the misclassification cost matrix (in the case of four categories):

$$
R_{ij} = \begin{bmatrix} 0 & u_{11} & u_{21} & u_{31} \\ o_{11} & 0 & u_{12} & u_{22} \\ o_{21} & o_{12} & 0 & u_{13} \\ o_{31} & o_{22} & o_{13} & 0 \end{bmatrix} \tag{3.1}
$$

With the constraints $u_{11} \leq u_{21} \leq u_{31} \geq u_{22} \geq u_{13}$, $u_{21} \geq u_{12} \leq u_{22}$, $o_{11} \leq o_{21} \leq o_{31} \geq o_{22} \geq o_{13}$, $o_{21} \geq o_{12} \leq o_{22}$. The misclassification cost matrix is (usually) symmetric and positive. The main diagonal is zero.

Additional Error Component

Especially for censored data, it seems plausible to assume a model like

$$
\tilde{Y} = \bar{Y}_k + \epsilon_k \text{ for } Y = k , \tag{3.2}
$$

where $\bar{Y}_k$ stores the average value of the class k and ϵ denotes the (class specific) error function, which describes the error made by grouping the values of the class. Depending on the choice of the error distribution of the regression model

ϵ and the class specific error functions ϵ_k, the model may not be identifiable[46]. In both cases, an Expectation Maximization (EM)-like algorithm has to be used.

In general, this approach of modeling ordinal data is used in analysis of censored data, as we have to specify the distributions with a known support. Nevertheless, we can use this concept in a slightly revised variant: We assume the k categories to be numbered subsequently with $\{1, \ldots, k\}$. Now, we arbitrarily set the midpoints between the numbers as class borders. For each of the supports $\{(-\infty, 0.5], (0.5, 1.5], \ldots, (k + 0.5, \infty)$ we have to specify a probability density function. Popular alternatives are:

- continuous uniform distribution
  The most plausible choice is the continuous uniform distribution as it reflects the missing knowledge best.

- (continuous) triangular distribution
  If the categories are not equally filled with objects, we might use a (generalized) triangular distribution. Is is called generalized because it does not have to be symmetric both in decrease to the tails and the value at the borders of the support.

- truncated Gaussian distribution
  The Gaussian distribution is not a good choice because of accumulative errors which would imply a central limit theorem to hold. That is not plausible, as the categories are constructed around a naturally occurring center and the variables are measured with error[47]. Often the number of categories is not really motivated by the task.

Only the approach with continuous uniform distributions seems to be usable for the intermediate categories. Unfortunately the two categories at the borders are in need of asymmetric distributions with an infinite support.

Scoring

If the data has been measured on an ordinal scale with a latent underlying variable, the scaling is not unique because every data source (participant) has

---

[46]If there are parameters for the class wise error functions left, even the parameter may not be identifiable.

[47]That means the expectation is the correct value.

got its own perception of the scale. For using this kind of ordinal data in analysis there has to be a kind of standardization of the perception. This may be done by collecting more than one grade for the same object by different individuals and by assuming a finite number of latent clusters of individuals having the same perception of scale in all clusters.

Unfortunately, the term "scoring" is not unique. While it is necessary to deal with latent ordinal data, it is often used in a more "direct" manner as the categories are assigned a number and treated as metric data. In this case "scoring" is the method of assigning the "right" numbers to the categories (Williams and Grizzle, 1972; Agresti, 1984; Fielding, 1997, 1999). For nonlinear regression functions, it is indeed possible to use fixed scores in regression as the different distances may be caught by the nonlinear function.[48]

Torra, Domingo-Ferrer, Mateo-Sanz and Ng (2006) suggest mapping an ordinal variable X to the unit interval ($[0, 1]$) with the help of a strict isotone mapping function $f : \{1, \ldots, k\} \rightarrow [0, 1]$. This approach is very similar to the one proposed for use in MLP networks (cf. subsection 4.6.1). The "right" mapping function has to be estimated together with the analysis function (Torra, Domingo-Ferrer, Mateo-Sanz and Ng, 2006, p. 469f.).

3.1 Analysis of Ordinal Input Data

Let f denote the regression function of some explanatory variables x on the response variables Y; flexibility results from a parameter vector θ.

$$Y = f(x|\theta) + \epsilon \; ,$$

where ϵ denotes an unbiased, homoscedastic and uncorrelated error. The usual approach for minimizing this function is the LS approach (cf. equation (2.5)).

For ordinal input data, it is common to assume an underlying continuous variable and to model this situation with the help of error components (Kukuk, 2002; Terza, 1987; Johnson, 2006). Unfortunately, this implies making parametric distributional assumptions on the errors. Furthermore, in the case of unknown thresholds the used generalized linear mixed model is not identifiable (Johnson, 2006, p. 288). This concept seems to be useful only for censored

---

[48]Scharfstein, Liang, Eaton and Chen (2001) provide such a method by using quadratic regression functions.

ordinal data with known underlying distribution (e.g. income groups) and thus is not further covered in this work.

Obviously, for grouped data there is no approach using the underlying variable assumption in an appropriate way. Therefore, we will have to use a more general approach. There are basically two methods for modeling ordinal input data without assuming an underlying latent variable; both are based on associating ordered "weights" to the different ordered categories. The two methods may be classified as parametric and nonparametric ones.

3.1.1 Choosing an Appropriate Mapping

Torra, Domingo-Ferrer, Mateo-Sanz and Ng (2006) suggest choosing a mapping function $m_o(\cdot)$ for every ordinal input variable $m_o : \{1, \ldots, k_o\} \rightarrow [0; 1]$. The functions $m_o(\cdot)$ have to be monotonic, but the identities $m_o(1) = 0$ and $m_o(k_o) = 1$ are not stated because every ordinal variable should have its own subdomain in $[0, 1]$ (Torra, Domingo-Ferrer, Mateo-Sanz and Ng, 2006, p. 470). They suggest to use a genetic algorithm in order to find the "right" set of mapping functions m_o for every ordinal variable within the model(Torra, Domingo-Ferrer, Mateo-Sanz and Ng, 2006, p. 470f.).

This approach is of nonparametric nature, but using a kind of "mapping" to the unit interval is appealing and may also be done with the help of a parametric approach with constraints as it is introduced in the next subsection.

3.1.2 Choosing an Appropriate Design Matrix

Let us assume, we had a regression model with o ordinal variables with s_i categories within the vector $\boldsymbol{x}$. The remaining variables measured on a metric scale contribute to the regression function directly with their values and the ones measured on a nominal scale are coded with a dummy or effect coding scheme resulting into $\sum_j k_j - 1$ binary values, where k_j denotes the corresponding number of categories (Judge, Griffiths, Carter Hill, Lütkepohl and Lee, 1985). There are different proposed methods for ordinal regressors depending on the kind of ordinal data to be modeled.

One of the most popular models in the case of censored ordinal data is to assume a rough measurement of the underlying metric variable and to estimate the expectation for the underlying variable $\tilde{X}$ depending on the measured realization of x $\mathbb{E}(\tilde{X}|x)$ (Ronning and Kukuk, 1996). This is only possible by assuming distributions for ϵ, and *all* regressors[49] $\boldsymbol{x}$ (Kukuk, 2002, p. 384); this model has been called *additional error component* model. By using simulated latent variables this approach does not need further assumptions on ϵ and other (metric and nominal) regressors than the ordinal ones (Breslaw and McIntosh, 1998). This approach requires distributional assumptions for the regressors which often lack reasoning. This is why this model is not taken into consideration in the subsequent parts of this thesis.

The following coding scheme has been proposed for simple ordinal data (Boyle, 1970; Walter, Feinstein and Wells, 1987): Each ordinal variable is replaced by a binary vector with size $s_i - 1$, thus the vector has one dimension less than there are categories. The vector is filled with 1 up to the position associated with the value of the ordinal variable. For example: Suppose an ordinal variable with three categories. This will result in a two dimensional vector and the three possible vectors are $(0,0)$, $(1,0)$ and $(1,1)$. While this coding scheme allows a simple interpretation of the corresponding (linear) parameter, it does not result in different models to the nominal coding schemes (MacDonald, 1973). We have to add constraints to make sure that the influence of the ordinal variable is monotone. This model will be able to estimate the *unknown misclassification costs* as it renders "utilities" that can be realized for reaching the next higher class. The ordinal structure of the variable is taken into account by using constraints. Alternative coding schemes, like the coding $(0,0)$, $(1,0)$, $(0,1)$ for three categories provide another interpretation of the ordinal variable. In the sequential coding scheme $(0,0)$, $(1,0)$, $(1,1)$, the higher classes include the lower ones, i.e. in estimation the higher classes also influence the estimators belonging to the lower classes. This is useful for sequential ordinal data like level of education, but not for simple ordinal data. In this case, a simple nominal coding scheme like $(0,0)$, $(1,0)$, $(0,1)$ with constraints is more appropriate because the higher classes do not include the lower ones.

---

[49]The Gaussian distribution is used for ordinal variables within $\boldsymbol{x}$ (Hsiao and Mountain, 1985).

Adding constraints can be done with the following constraints on the parameters $\beta_2 \ldots \beta_s$ corresponding to a single ordinal variable:

$$\beta_r = \alpha \cdot \sum_{i=1}^{r-1} \alpha_i \text{ with } 0 < \alpha_i < 1 \quad \forall i \; ,$$

$$\sum_{i=2}^{s} \alpha_i = 1$$

(3.3)

The restrictions on α_i ensure that the monotonicity is given and the α is needed to discriminate between a monotone increasing and decreasing relationship. Because of identifiability of the parameters the normalization restriction and the convention $\beta_1 = 0$ is essential. As the value of α can be interpreted as the influence of the ordinal variable on the outcome, we will reference it as θ in analogue to metric explanatory variables. The same test procedures may be used for it, i.e. the t-test for significance of the variable.

The constraints are not used in good-natured problems, in which the ordering is supported by the data. We will get a solution equivalent to the standard covariance analysis, but with parameter estimates, that are more easy to interpret.

Because this coding is the ordinal analogue to the nominal dummy coding, it suffers from the same interpretation problem on the constant factor for linear models as the lowest category is associated with the constant factor. If the constant factor should describe the "mean" effect, the nominal effect coding can be made ordinal using the following codings: $(-1, 0)$, $(0, 0)$, $(0, 1)$ for three categories. The effect coding does not make sense for an even number of categories, because there is no explicit middle class.

Now, we have to solve the problem in general

$$\text{Minimize} \quad \frac{1}{2} \sum_{n} (y_n - f(\boldsymbol{x}_n | \boldsymbol{\theta}))^2$$

$$\text{subject to} \quad -\alpha_{ij} < 0 \quad \forall i, j$$

$$\sum_{j}^{s_i} \alpha_{ij} - 1 = 0 \quad \forall i$$

(3.4)

where α_{ij} denotes the influence of stage j of the ordinal variable i on the response variable.

The use of a quadratic target function with constraints suggests using quadratic programming (Bazaraa, Sherali and Shetty, 1993, p. 503ff.). However, the parameters belonging to the ordinal variables are nonlinear and the problem (3.4) has to be solved by using barrier functions (Bazaraa, Sherali and Shetty, 1993)

$$\text{Minimize} \quad \inf\left\{\frac{1}{2}\sum_n (y_n - f(\boldsymbol{x}_n|\boldsymbol{\theta}))^2 + \mu B(\boldsymbol{\alpha}) : \alpha_{ij} > 0 \quad \forall i, j)\right\}$$

$$\text{subject to} \quad \mu > 0 \tag{3.5}$$

$$\sum_j^{s_i} \alpha_{ij} - 1 = 0 \quad \forall i \ ,$$

with $B(\boldsymbol{t})$ denoting Frisch's logarithmic barrier function. Here we have

$$B(\boldsymbol{\alpha}) = -\sum_{i,j} \log[\alpha_{i,j}] \ .$$

In general, the problem (3.5) is solved with the following two-step procedure (Bazaraa, Sherali and Shetty, 1993, p. 391):

Algorithm 3.1

Init   *Let $\epsilon > 0$ be a termination scalar, and choose a point $\boldsymbol{\alpha}^{(0)}$ within the admissible region. Let $\mu^{(0)} > 0$ and $k = 0$.*

Main   1. *Starting with weights in iteration k $\boldsymbol{\alpha}^{(k)}$, solve (3.5) with μ replaced by $\mu^{(k)}$ with the help of an unconstraint optimization algorithm. Let $\boldsymbol{\alpha}^{(k+1)}$ be the optimal solution.*

2. *If $\mu^{(k)} B(\boldsymbol{\alpha}^{(k+1)}) < \epsilon$, stop. Otherwise reduce μ_k to μ_{k+1} and repeat step 1.*

The remaining constraint in (3.3) $\sum_j^{s_i} \alpha_{ij} - 1 = 0$ can be solved by Lagrange optimization (Bazaraa, Sherali and Shetty, 1993). This implies that the involved derivatives are no longer independent from each other and that the parameters α_{ij} for a fixed i have to be estimated simultaneously.

A possible starting value for the parameter vector of the nonlinear regression can be obtained as follows: Fix the values $\alpha_{i,j}$ by using the equal distance assumption, i.e. setting $\alpha i, j = \frac{j}{s_i}$ for all i and $j \in \{1, \ldots, s_i\}$. Reduce the design matrix, so that for each ordinal variable only one column is left. Now, we have to solve an unconstraint linear Regression problem. The estimated parameters for α and β are used as starting values for algorithm 3.1, whereas the constant term has to be increased by the default effect left out of the full design matrix.

Modeling Errors in Regressors

In particular, models need a component for describing the error of misclassi-fication for ordinal data scales, for which no general perception of the scale exists. We can expect that the differences in perception of the scale for semi-standardized models are restricted to categories in the close neighborhood of the current value. In contrast, we have to model this influence for the individual scale. These models are very over-parameterized and only data analysis with repeated measurements are possible.

Therefore, we usually have to expect a semi-standardized level of perception of the scale, where we can reasonably assume, that the errors are neglectable in size and are sufficiently covered by the error of the model.

Symmetric Ordinal Data

A special kind of ordinal data, called symmetric ordinal data, has been in-troduced in subsection 2.1.4. This definition is important as the number of parameters needed for the model can be reduced significantly. This is pos-sible because we can use an effect-coding scheme: For an uneven number of categories c we need a coding vector of length $\frac{c-1}{2}$ and we can code the first half by -1 and the second by 1: For example, 5 categories can be coded as $(-1, -1)$, $(0, -1)$, $(0, 0)$, $(0, 1)$ and $(1, 1)$. Obviously, the class in the middle is the reference class and the neighboring categories have equal distances. Symmetric coding with only three classes is equivalent to assuming a metric variable because the categories are equidistant. Using this coding scheme, the monotonicity is only partially satisfied among the categories that are mirrored. The monotonicity has to be ensured with the help of restrictions, too.

Strictly speaking, additional constraints like these are the reasons why ordinal explanatory variables should be treated differently from nominal data. As already mentioned, the results are equivalent in the normal case. Any monotone relationship may be represented in linear regression with ordinal input values and linearity can be obtained by relocating the ordinal input variable. In

Figure 3.1 this is demonstrated with the help of a nonlinear relationship $(f(x) = 4 \cdot \exp(2 \cdot x) - 4)$, which has been the basis for a simulation of metric responses to a 4 categories variable with the "real"—but unobserved—values $-0.75, -0.25, 0.25$ and 0.75. The values -0.75, ca. -0.6, ca. -0.25 and 0.75 are associated to the four categories to obtain a good linear relationship. Therefore, by associating other values to the categories of the latent explanatory variable a perfect linear relationship can be obtained.

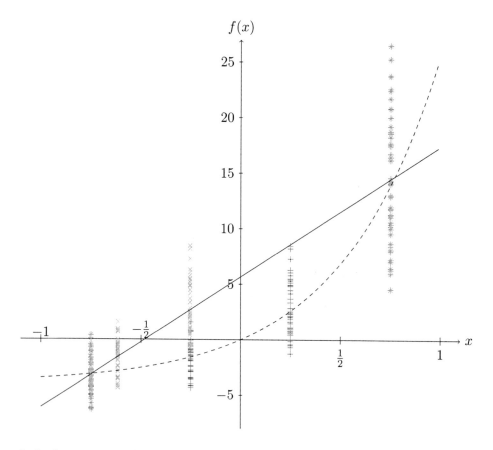

dashed: monotone function, solid: linear regression with arbitrary location of classes

+: grouped simulated data, x: for regression relocated data

Fig. 3.1: Grouped monotone function to linear regression

3.2 General Approaches to Ordinal Output Data Modeling

Trying to construct a model for ordinal response variables the first and most intuitive idea is to use subsequent binary classifications (Marshall, 1999; Frank and Hall, 2001). An ordinal response variable model with k categories is composed by using $k - 1$ binary models distinguishing lower versus higher categories between each of the categories. In some models, not *all* higher or lower categories are taken into account. Although this is the basic idea behind many of the concrete models, there are also approaches based on variants of the multinomial distribution and/or by using constraints.

Some methods, that are important for ordinal input data modeling like the mapping by Torra, Domingo-Ferrer, Mateo-Sanz and Ng (2006), are used for ordinal output data modeling, too.

3.2.1 Properties for Ordinal Regression Models

We should state that the decision boundaries in the feature space

1) should not cross in the "interesting part" of the feature space,

2) should be ordered as like as the categories.

That means, for every point x in the part of the feature space, that is used for prediction, holds: $\exists_{\epsilon > 0} \#\{f(y) : d(x,y) < \epsilon\} \leq 2$ and the two categories (at maximum in this set) are neighbored according to the ordering of the response variable's values.

The proportional odds assumption[50] is a popular approach to ensure the first property in ordered regression with linear boundaries as it ensures parallel decision boundaries (McCullagh, 1980). This is very restrictive as the non-crossing behavior is given for the complete space and not only for the used part of it. Therefore, there are approaches to relax the strong proportional odds assumption (Peterson and Harrell, 1990). In general, i.e. without the proportional odds assumption, the first property is not easy to obtain in nonlinear regression. If the first property is given, the second property can be satisfied by using ordered constants as biases within the models.

---

[50]see 3.3.1

Before we present some classical approaches, we have to formulate some plausible criteria for an appropriate ordinal modeling of output variables additionally to the axioms of section 2.1.1:

1) The a-posteriori probability should be unimodal, i.e. the probability is monotonely decreasing to the flanks. Neighboring categories may share the modal value. This property is automatically obtained, if the decision boundaries are ordered and non-crossing.

2) Reversibility, i.e. the decision boundaries are invariant from changing the ordering direction of the response variable. This property is superfluous, if the ordering of the categories is naturally given (e.g. sequential models).

3) Stochastic orderings are recommended for linear models, i.e. for two realizations of the predictor x_d, $x_{d1} < x_{d2}$ with $\beta_d > 0$ we would expect the outcome distribution $(p_{11}, \ldots, p_{k1})$ for x_{d1} to be stochastically larger than $(p_{12}, \ldots, p_{k2})$ for x_{d2}.

In practice, we have to know which sampling methods can be used with the intended model. These are the most common random sampling methods (Cochran, 1977):

- The simple random sampling collects data in such a way that every sample has an equal chance of being selected from the population.

- The stratified random sampling collects samples from different strata to ensure (especially for less occurring combinations) a nearly complete coverage of the interesting space. Concerning input-output models, we can distinguish between input-stratified and output-stratified sampling.

- The cluster sampling is done by selecting homogenous groups of samples. For example, if we would like to analyze the effect of living conditions on the inhabitants satisfaction, we could choose small random parts of cities (usually having the same or nearly equivalent living conditions) and interview residents about their satisfaction.

Fixing the predictors does not change their influence on the regression and that is the reason why input-stratified and cluster sampling are unproblematic. Nevertheless, the output-stratified sampling may cause problems as the direction of the influence is "inverted", i.e. we do not have the correct relation between the different outcomes as they are fixed externally.

For further comparison of the different methods introduced in the subsequent parts of this section we are in need of an unified view on these techniques. Li and Lin (2006) and Cardoso and da Costa (2007) developed equivalent unified models with slightly different formalizations (Cardoso and da Costa, 2007; Li and Lin, 2006). While the work by Cardoso and da Costa does not provide any enlightening results, the model by Li and Lin can help to understand the ordinal classification models better.

The framework for ordinal classification developed by Li and Lin is well suited to analyze ordinal classification models. It is based on a so called V-shaped cost matrix (Li and Lin, 2006, p. 866):

Definition 3.1 (V-shaped cost matrix)
A cost matrix C is called V-shaped, if

$$C_{i,j-1} \geq C_{i,j} \text{ if } j \leq i \text{ and}$$
$$C_{i,j} \leq C_{i,j+1} \text{ if } j \geq i \ .$$

It is assumed for ordinal classification, that each row of the V-shaped cost matrix is convex:

Definition 3.2 (convex rows of a matrix)
A row of a cost matrix C is called convex, if for $1 < j < k$

$$C_{i,j+1} - C_{i,j} \geq C_{i,j} - C_{i,j-1} \ .$$

Using a threshold model for classification of the form $f(\boldsymbol{x}) - \theta_r$ the following fact can be proved (Li and Lin, 2006, p. 868):

Theorem 3.3 (ordered thresholds)
If every row of the cost matrix is convex, and the binary classification algorithm minimized the loss

$$\Lambda(f) + \sum_{i=1}^{n} \sum_{r=1}^{k-1} |C_{y_i,r} - C_{y_i,r+1}| \cdot \ell \left(y_i^{(r)}(f(\boldsymbol{x}_i) - \theta_r) \right) \tag{3.6}$$

where $y_i^{(r)} = 2\mathbb{1}\{r < y_i\} - 1$ and $\ell(\cdot)$ is non-increasing, there exists an optimal solution $(f^*, \boldsymbol{\theta}^*)$ such that $\boldsymbol{\theta}^*$ is ordered. $\Lambda(f)$ denotes the parts of the optimization which are only related to f, i.e. in the parametric case to the parameters of the function, but neither to $\boldsymbol{\theta}$ nor any element of the sample or the cost matrix.

Comment 3.4

This theorem seems to be appealing as it states, that additional constraints on the thresholds $\boldsymbol{\theta}$ are superfluous. Unfortunately, we need to know the cost matrix in order to make use of Theorem 3.3. Therefore this theorem only transfers the problem to another place without really solving it.

3.2.2 Approaches to Ordinal Output Data Modeling

Because there are different kinds of ordinal data (cf. subsection 2.1.4), we have to construct appropriate models according to each of these different settings in order to use the maximum possible amount of information. Table 3.1 provides an overview of the different approaches introduced briefly in this subsection.

Approach	Type of Ordinal Data	Models
median classification	grouped	(Lee, 1992)
underlying metric variable	grouped	cumulative
sequential modeling	sequential	sequential model, continuation-ratio
errors on variables	latent (grouped/sequential)	see literature given in text
scoring	latent (grouped/sequential)	see literature given in text
set of binary splits	grouped/simple ordinal	see literature given in text
multinomial dist. with costs	simple ordinal	(Kotsiantis and Pintelas, 2004)
constraint multinomial dist.	simple ordinal	stereotype, adjacent-category
ordered multinomial dist.	simple ordinal	direct applicable

Tab. 3.1: Approaches to Ordinal Data Modeling

Barnhart and Sampson (1994) distinguish four basically different models for ordinal output (logit or hazard) data analysis:[51]

(i) adjacent-categories

(ii) cumulative odds

---

[51] The introduced terms are related to models, which can be found in sections 3.3 and 3.4.

(iii) forward continuation-ratio

(iv) backward continuation-ratio

They did not mention the stereotype and the sequential model.

Median Classification

A special form of regression on an ordinal variable with an underlying metric variable has been proposed under the label "median regression" by (Lee, 1992). The characteristic idea of the median regression is that there is a link between the median of the output and the regressors

$$\text{Med}(Y|\boldsymbol{x}) = f(\boldsymbol{x}) \ . \tag{3.7}$$

The discrete output is generated by estimating appropriate cut points for the categories.

Similar to ordinary regression, this model makes no particular assumptions on the errors except that is has to be unbiased and homoscedastic.

Although the basic idea is similar to ordinary regression, it should be called median classification as it tries to classify directly without using any regression on parameters of a probability function. This approach to ordinal data analysis is not presented in detail as MLP are regression models.

Multinomial Distribution Model with Cost Information

Another model which is as intuitive as the median regression/classification but closer to nominal data analysis is based on cost information. Unfortunately, the cost information is often unknown. Therefore, some authors proposed the simple misclassification cost matrix

$$C = \begin{bmatrix} 0 & 1 & 2 & \dots & k-1 \\ 1 & 0 & 1 & \dots & k-2 \\ \vdots & & \ddots & & \vdots \\ k-2 & \dots & 1 & 0 & 1 \\ k-1 & \dots & 2 & 1 & 0 \end{bmatrix}$$

implicitly assuming equal "cost distances" between the categories (Kotsiantis and Pintelas, 2004). Symmetry of the matrix C is plausible and adding up costs of neighboring classes to obtain the costs for larger "differences" is plausible.

Although it is possible to reduce the number of unknown parameters of the cost matrix by additional assumptions on the data scale—like symmetry—it is not possible to define a cost matrix completely without any (maybe only partly-known) cost information.

Models with Underlying Metric Variables

Often ordinal values are generated by cutting a continuous variable into intervals (of different length) without knowing the exact borders. A natural way to model this kind of data is by using an underlying metric variable with additional parameters describing the borders of the categories. This kind of Models is analyzed in subsection 3.3.

Set of Binary Splits

This simple and intuitive approach has already been mentioned at the beginning of this subsection. It is intuitive as we only have to render decision boundaries to distinguish "higher" and "lower" objects. In addition, this meta analysis can be implemented in an easy way, because it reuses ordinary binary classification and regression techniques (Marshall, 1999; Frank and Hall, 2001). For example, Cheng (2007) used this basic idea for constructing a MLP network for ordinal output data. "Higher" and "lower" categories are distinguished in its basic form. Obviously, we assume a certain amount of homogeneity within the "higher" and the "lower" category independently from the current combination. For example with three categories, class one is separated from classes two and three at first and the first two classes are separated from class three afterwards. Therefore, class two is not too different from class one nor class three. If the categories are too different to be putted together in groups, we can build up $k - 1$ binary classifications between neighboring classes and ignore the rest of the sample (cf. i.e. adjacent-categories model in section 3.4).

Sequential Model

There are a lot of ordinal data which imply that an object falling into category r has been in category $r - 1$ before (Tutz, 1990, 1991). Therefore, it is plausible to model the probability of staying in category r conditional to having already achieved this category $P(Y = r | Y \geq r, \boldsymbol{x})$.[52] Further analysis of this model is done in subsection 3.3.2.

---

[52] Another interpretation is given by McCullagh and Nelder (1989), who have analyzed these models at first and calling them hierarchical or nested models (cf. subsection 2.1.4).

Correction for Models with Errors on the Latent Underlying Variable

If the underlying variable is latent, we have to expect, that individual influence on the scale is given. One possibility of facing this problem is to allow individual offsets on the scale by adding a random shift to the thresholds but holding their distances stable (Farewell, 1982). This approach reduces inter-subject differences in "leveling" of the scale but without adjusting its range. A more general case of random effect modeling for ordinal data in regression models is provided by Tutz and Hennevogl (1996) using alternative link functions and an EM algorithm for estimation. The model has the general form

$$P(Y_{it} \leq r|x_{it}, b_i) = F(z'_{itr}\beta + w'_{itr}b_i) \ ,$$

where z_{itr}, w_{itr} are design vectors composed from x_{it} for observation t in cluster[53], β is the fixed parameter and b_i are random effects with $b_i \sim \mathcal{N}(0, Q)$ for some covariance matrix Q (Tutz and Hennevogl, 1996, p. 542). Any more complex formulation would not be identifiable.

Scoring for Output Data

Scoring is an approach based on an additional error component by assuming one parametric distribution $P(\cdot)$ of the values on the latent scale for each of these clusters. Snell (1964) proposes the logit distribution for the probability of an observation of cluster i in category s_i with boundaries x_i and x_{i-1}

$$P_i(x_i) - P_i(x_{i-1}), \quad i = 1, 2, \ldots, m; j = 1, 2, \ldots, k \ . \tag{3.8}$$

An EM algorithm is needed to provide estimators for the boundaries x_i on this scale as the clusters within the individuals are unknown (Snell, 1964). The attached scores for each of the categories are calculated as midpoints between the boundaries. Other scoring methods and approaches to these random-effects models are discussed by Fielding (1997) and Fielding (1999).

Ordered Multinomial Distribution Model

There are a lot of distributions for interval scaled data—like normal, Poisson or exponential distribution—but just one for categorical data: The multino-

---

[53]Subjects with the same perception. Usually the same person.

mial distribution. Now the question arises, if there is a special multinomial distribution that is able to reflect the order structure of the data.

In natural parametrization (i.e. $\theta_i = \log \frac{p_i}{1-p_i}$), the complete ordered multinomial distribution for k categories looks like

$$f_{\boldsymbol{p}}(\boldsymbol{x}) = h(\boldsymbol{x}) \exp \left[\sum_{i=1}^{k} s_i \theta_i - n \ln \left\{ \sum_{j=1}^{k} \exp \left(\sum_{i=1}^{j} \theta_i \right) \right\} \right] \quad (3.9)$$

while

$$\boldsymbol{\theta} = (\theta_1, \ldots, \theta_k) = \left(\log p_1, \log \frac{p_2}{p_1}, \ldots, \log \frac{p_k}{p_{k-1}} \right)^t$$

$$p_j = \frac{\exp(\theta_1 + \cdots + \theta_j)}{\sum_{j=1}^{k} \exp \left(\sum_{i=1}^{j} \theta_i \right)} \quad \text{and} \quad s_j = \sum_{i=j}^{k} x_i$$

and x_i denoting the number of objects in category i (Wang, 1986). It is seen that $\boldsymbol{\theta}$ is the vector of the natural parameter and $\boldsymbol{s}$ is the order-dependent sufficient statistics for $\boldsymbol{\theta}$ (Wang, 1986).

As the order-dependent multinomial model by Wang is a distribution, we could extend it by introducing the following constraint:

$$\theta_1 \geq \theta_2 \geq \ldots \geq \theta_k \quad (3.10)$$

This restriction leads to a concave shape of the response distribution.

Then we can state:

Theorem 3.5
Under using constraint (3.10), the parametrization (3.9) is unimodal.

Proof. Let us assume, there were $m < n < \ell$ with $p_m > p_n < p_\ell$. This would imply:

$$\sum_{i=1}^{n} \theta_i < \sum_{i=1}^{m} \theta_i \Rightarrow \sum_{i=m+1}^{n} \theta_i < 0 \text{ and}$$

$$\sum_{i=1}^{n} \theta_i < \sum_{i=1}^{\ell} \theta_i \Rightarrow \sum_{i=n+1}^{\ell} \theta_i > 0$$

With $\min_{i \in \{m+1,\ldots,n\}} \{\theta_i\} \geq \max_{i \in \{n+1,\ldots,\ell\}} \{\theta_i\}$ we have a contradiction. □

Using the transformations

$$\theta_i = 1 - \sum_{j=1}^{k-1} \Delta_j \text{ and } t_i = \sum_{j=i}^{k} s_j = \sum_{j=i}^{k} j \cdot x_j \qquad (3.11)$$

it is easily seen that Δ_i are the parameters to the ordered and concave multinomial distribution (3.9) with statistics t. Then the class probabilities are obtained by

$$p_j = \frac{\exp(j)\exp(-\sum_{i=1}^{j-1}\Delta_i)}{\sum_{j=1}^{k}\exp(j)\exp\left(-\sum_{i=1}^{j-1}\Delta_i\right)} \qquad (3.12)$$

Symmetric Ordinal Data

Concerning symmetric ordinal data as output, we have to rebuild the polytomous categorical regression from (2.13) by replacing the category-specific offsets β_{0r} in the following way (for an uneven number of categories):

$$\beta'_{0,1} = \beta_0 - \beta_{0,1}$$

$$\vdots \qquad \vdots$$

$$\beta'_{0,\frac{k-1}{2}} = \beta_0 - \beta_{0,\frac{k-1}{2}}$$

$$\beta'_{0,\frac{k+1}{2}+1} = \beta_0 + \beta_{0,\frac{k-1}{2}} \qquad (3.13)$$

$$\vdots \qquad \vdots$$

$$\beta'_{0,k} = \beta_0 + \beta_{0,1}$$

β_0 is unconstraint, but constraints are necessary to ensure $\beta_{0,1} > \beta_{0,2} > \cdots > \beta_{0,\frac{k-1}{2}} > 0$. The middle category is left out as it is determined by the others. Thus, the number of parameters is reduced from $k - 1$ to $\frac{k}{2}$. Unfortunately, this scheme is not applicable for even numbers of categories because we have to leave out one of the categories as the redundant one and are left with an uneven number of regressions which cannot be treated symmetrically.

3.2.3 Non- and Semiparametric Analysis of Ordinal Output Data

Besides the parametric classification models for ordinal data, there are also some approaches to non or semiparametric classification of ordinal data. The most popular and researched nonparametric approach is SVM for ordinal data which are introduced in the following paragraphs. Ordinal data trees based on CART are also introduced briefly.

There are ordinal approaches to kernel regression (Kauermann, 2000) and ordinal additive models with smoothed one-dimensional functions for each regressor (Hastie and Tibshirani, 1987) and some more, which are not covered here.

Concerning nonlinear models, there is no difference if ordinal data is based on any type of continuous underlying or not. Sequential and non-sequential ordinal data are not distinguished although this could lead to a better fit of the models as the models are more restricted.

The most challenging task in nonlinear ordinal regression or classification is to provide decision boundaries in the feature space, which do not cross and are ordered. We will describe two possible solutions to ensure these properties in SVM and in CART.

Support Vector Machines

Herbrich *et al.* provided one of the first approaches to ordinal SVM classification (Cristianini and Shawe-Taylor, 2000) for ordinal data with a metric underlying (Herbrich, Graepel and Obermayer, 1999b,a). Their approach has been further (especially the algorithmic part) developed by others to provide a lower algorithmic complexity (Chu and Keerthi, 2005) or more sophisticated parallel boundaries with or without the same maximal margin between the different categories (Shashua and Levin, 2002). The basic idea in each approach is to use linear decision boundaries and ordering constraints within the induced kernel space; i.e. in the kernel space the decision boundaries are ordered and parallel. As Kernel functions provide a continuous projection of the feature space into the kernel space, the induced decision boundaries in the feature space are also ordered and free of crossings.

Trees

The first approaches to ordinal classification trees have been published by Potharst and Bioch (1999, 2000). Unfortunately they assume that the explanatory variables are ordered as well. This implies a very strong restriction as nominal variables can not be used. Therefore Kramer *et al.* provide an approach based on Structured Regression Tree and Classification Tree (S-CART) (Kramer, 1996). They try to use a kind of scoring to represent the ordinal structure and use a metric scaled measure of variances for their splitting rule (Kramer, Widmer, Pfahringer and de Groeve, 2000). Another approach, which is completely different from the former one, is based on ordinal impurity measures on scores by using the median of the nodes (Piccarreta, 2004). It has to be mentioned, that this also is no real ordinal measurement of impurity. The most complex approach has been proposed by Cao-Van and De Baets (2003) and is based on the multi-criteria decision aid and combinations or ordered relationships to construct partial dominance. It is very different to the original CART algorithms by Breiman, Friedman, Olshen and Stone (1984).

3.3 Models for Ordinal Data with an Underlying Metric Variable

The models introduced in this and the following sections are all based on a mapping of $\mathbb{R}$ to the interval $[0, 1]$ with the help of a link function $F(\cdot)$. This function is often related to the assumed error distribution on the underlying variables of the ordinal response variable (Boes and Winkelmann, 2006). Among others, the following four alternatives for $F(\cdot)$ are the most popular ones (Ananth and Kleinbaum, 1997):(McCullagh and Nelder, 1989, p. 30)

$$F(t) = \frac{\exp t}{1 + \exp t} \qquad \text{logit}$$

$$F(t) = \Phi(t) \qquad \text{probit}$$

$$F(t) = 1 - \exp\left(-\exp(t)\right) \qquad \text{log-log}$$

$$F(t) = \tan\left(\pi \cdot \left(t - \frac{1}{2}\right)\right) \qquad \text{inverse Cauchy}$$

As the choice od $F(\cdot)$ depends on the application. The models are treated in their general form here.

3.3.1 Cumulative Model

The most natural way to define a model for ordinal data with an underlying metric variable is to use the cumulative model (Ananth and Kleinbaum, 1997; Anderson and Philips, 1981; Liu and Agresti, 2005)

$$P(Y \leq r|\boldsymbol{x}) = F(\theta_r + f(\boldsymbol{x})) \quad (r = 1, \ldots, k - 1) \tag{3.14}$$

by modeling the probability that the output Y supersedes a specific class r with the help of a (non-)parametric function f of the regressors $\boldsymbol{x}$.

Assuming a metric variable $\tilde{Y}$ which has been divided into k intervals to form the different ordered categories, the probability for class r is obtained by

$$P(Y = r|\boldsymbol{x}) = P(\theta_{r-1} < \tilde{Y} \leq \theta_r|\boldsymbol{x}) \ .$$

If we further assume an additive unbiased error model for $\tilde{Y}$

$$\tilde{Y} = f(\boldsymbol{x}) + \epsilon$$

with $F(\cdot)$ denoting the CDF of ϵ. Then it is easily proved that

$$P(Y = r|\boldsymbol{x}) = F(\theta_r + f(\boldsymbol{x})) - F(\theta_{r-1} + f(\boldsymbol{x})) \ . \tag{3.15}$$

This is the basic cumulative model. Using a linear function $f(\boldsymbol{x}|\boldsymbol{\beta}) = (1, \boldsymbol{x}^t)\boldsymbol{\beta}$ we obtain the proportional-odds model. To ensure the ordering, the constant parameters $\beta_{0,1}, \ldots, \beta_{0,k-1}$ within cumulative models have to be constraint by $\beta_{0,1} < \cdots < \beta_{0,k-1}$ and one of them has to be fixed for the sake of identifiability.

Depending on the choice of the link function $F^{-1}(\cdot)$ cumulative odds models are invariant to a reverse of the categories (McCullagh, 1980, p. 116). The use of the logistic, probit or inverse Cauchy provides reversibility. This is not the case for the log-log and the complementary log-log (cf. subsection 3.3.3). Cumulative odds models are sensitive to the sampling fractions of the output Y for the categories $1, \ldots, k$, i.e. output-stratified sampling—like it is used in case-control studies—is not applicable in contrast to the ordinary logistic regression (Greenland, 1994, p. 1997).

Estimation

Linear cumulative odds models can be rewritten to

$$P(Y = r|\boldsymbol{x}) = F(\beta_{0,r} + \langle \boldsymbol{x}, \boldsymbol{\beta_r^-} \rangle) - F(\beta_{0,r-1} + \langle \boldsymbol{x}, \boldsymbol{\beta_r^-} \rangle) \ , \qquad (3.16)$$

i.e. with the special link function

$$g_r(\pi_1, \ldots, \pi_{k-1}) = F^{-1}(\pi_1 + \cdots + \pi_r) \quad \forall 1 \leq r \leq k-1$$

and restrictions $\beta_{0,1} < \cdots < \beta_{0,k-1}$. They can be estimated like any other GLM (Armstrong and Sloan, 1989). Fortunately, with the help of the following reparametrization

$$\delta_{0,1} = \beta_{0,1} \quad \text{and} \quad \delta_{0,r} = \log[\beta_{0,r} - \beta_{0,r-1}]$$

the model (3.16) may be written to

$$F^{-1}(P(Y = r|\boldsymbol{x})) = \delta_{0,1} + \langle \boldsymbol{x}, \boldsymbol{\beta^-} \rangle \ ,$$

which can be estimated without constraints.

Generalized Cumulative Model

The most general form—using a linear function—of the cumulative model is

$$P(Y \leq r|\boldsymbol{x}) = F(\beta_{0r} + \langle \boldsymbol{x}, \boldsymbol{\beta_r} \rangle) \quad (r = 1, \ldots, k-1) \ . \qquad (3.17)$$

This model is very important for testing the proportionality assumption used in many applications.

Using the idea of a location-scale model (3.23), we can extend the general model (3.17) to

$$P(Y \leq r|\boldsymbol{x}) = F\left(\frac{(\beta_{0r} + \langle \boldsymbol{x}, \boldsymbol{\beta_r} \rangle)}{\langle \boldsymbol{x}, \boldsymbol{\gamma_r} \rangle} \right) \quad (r = 1, \ldots, k-1) \ , \qquad (3.18)$$

where $\boldsymbol{\gamma_r} \in \mathbb{R}^d$ is an addition scaling vector.

Proportionality

Strictly speaking, the model is called proportional odds model, if the link function is the logistic function. It is called proportional hazards model, if it uses the complementary log-log function. Nevertheless, these models can (and have been) analyzed as realizations of the more general model (3.14) with a linear function $f(\cdot)$ (Walker and Duncan, 1967; McCullagh, 1980):

$$P(Y \leq r|\boldsymbol{x}) = F\big(\beta_{0,r} + \langle \boldsymbol{x}, \boldsymbol{\beta}^- \rangle\big) \quad (r = 1, \ldots, k-1) \tag{3.19}$$

The most eye-catching characteristic of this model is the fixed parameter $\boldsymbol{\beta}$ for all categories. This implies, that the odds of two different vectors $\boldsymbol{x}_1$, $\boldsymbol{x}_2$ in the case of a logistic function are modeled by

$$\frac{P(Y \leq r|\boldsymbol{x}_1)}{P(Y \leq r|\boldsymbol{x}_2)} = \exp\Big[\langle \boldsymbol{x}_1 - \boldsymbol{x}_2, \boldsymbol{\beta}^- \rangle\Big] \ .$$

That is the reason why the model is called proportional (McCullagh, 1980). Together with the constraints on the constant parameters, this feature leads to a set of ordered parallel linear decision boundaries within the feature space that do not cross. Models with the proportionality assumption are stochastically ordered (McCullagh, 1980; Barnhart and Sampson, 1994).

The proportionality assumption holds, if the regression coefficients are independent from r (McCullagh, 1980). Testing this property therefore has to incorporate the generalized cumulative model (3.17) with potentially unequal $\boldsymbol{\beta}$ for each category and testing the hypothesis: $\boldsymbol{\beta}_i = \boldsymbol{\beta}_j$ for all $i, j \in \{1, \ldots, k-1\}$. There are two approaches for performing this test problem in literature: The score test (Stiger, Barnhart and Williamson, 1999) and the more popular Wald test for proportionality (Peterson and Harrell, 1990; Brant, 1990). The method proposed by McCullagh (1980) uses another generalization of the model 3.19: A subpopulation-dependent scalar τ_i additionally allows variation in variance over the m subpopulations:

$$P(Y_i \leq r|\boldsymbol{x_i}) = F\left(\frac{\beta_{0,r} + \langle \boldsymbol{x_i}, \boldsymbol{\beta}^- \rangle}{\tau_i}\right) \quad (r = 1, \ldots, k-1) \tag{3.20}$$

Testing for $\tau_i = \tau_j$ for all $i, j \in \{1, \ldots, m\}$ also yields a test of the proportional odds assumption. Nevertheless, this model may be used independently from testing because it also describes a model for ordinal output data with an underlying continuous variable with a subpopulation-dependent error function following a scaled error function (Peterson and Harrell, 1990). Unfortunately, the subpopulations are usually not known a-priori and an EM algorithm is needed to distinguish the various sub-populations.

Assuming proportionality is directly related to assuming linearity of the model, i.e. if we can assume a linear relationship between two groups of continuous variables X and Y where the output variable Y is grouped, proportionality is given automatically.

Farewell (1982) has developed a test for the assumption that all observations share a common set of thresholds $\beta_{0,1}$ to $\beta_{0,k-1}$.

Partial Proportionality

The strong assumption of proportional odds restricts the flexibility of the model. Therefore, it can be relaxed by allowing more flexibility step by step. Peterson and Harrell (1990) introduced two groups of so called partial proportional models: Constraint and unconstraint partial proportional models, whereas the unconstraint model is the most flexible:

$$P(Y \leq r|\boldsymbol{x}) = F(-\beta_{0,r} - \langle \boldsymbol{x}, \boldsymbol{\beta}^- \rangle - \langle \boldsymbol{z}, \boldsymbol{\gamma}_r \rangle) \quad (r = 1, \ldots, k-1) , \quad (3.21)$$

where $\boldsymbol{z}$ denotes the subset of $\boldsymbol{x}$ for which the proportionality assumption does not hold and the corresponding parameters $\boldsymbol{\gamma}_r$ describe the deviation from the proportionality assumption. Thus, testing for $\boldsymbol{\gamma} = 0$ is a test for the proportionality assumption (Peterson and Harrell, 1990).

As this model includes a lot of additional parameters, the constraint partial proportionality models can be used, to prevent over-parametrization.

Peterson and Harrell (1990) introduced two constraint models. The first model is limited by a constraint on the $\boldsymbol{\gamma}$

$$P(Y \leq r|\boldsymbol{x}) = F(-\beta_{0,r} - \langle \boldsymbol{x}, \boldsymbol{\beta}^- \rangle - \langle \boldsymbol{z}, \boldsymbol{\gamma} \rangle \Gamma_r) \quad (r = 1, \ldots, k-1) , \quad (3.22)$$

with prespecified scalars Γ_r and $\Gamma_1 = 0$ for identifiability. This model reduces the number of additional parameters because the vector $\boldsymbol{\gamma}$ is no longer dependent on the category. Additionally, the specificity of every non-proportional predictor to the categories varies equally throughout the categories with Γ_r. Model (3.22) is appropriate, if linear variation of the $\boldsymbol{\gamma}$ parameters in r is assumed (Peterson and Harrell, 1990, p. 209).

A more flexible constraint model of (3.21) is

$$P(Y \leq r|\boldsymbol{x}) = F(-\beta_{0,r} - \langle \boldsymbol{x}, \boldsymbol{\beta}^- \rangle - \boldsymbol{z}^t(\mathrm{diag}(\boldsymbol{\Gamma}_r))\boldsymbol{\gamma}) \quad (r = 1, \ldots, k-1) \ ,$$

where $\boldsymbol{\Gamma}_r$ is a prespecified vector of as many entries as $\boldsymbol{\gamma}$, which are used with the corresponding predictor variable. That is why this model is less flexible than model (3.21) but more flexible than model (3.22).

These models can also be used for constructing a likelihood ratio or score test for proportionality (Peterson and Harrell, 1990).

Cox (1995) extended the unconstraint partial proportional odds model (3.21) by extending the location-scale model (3.20) to

$$P(Y \leq r|\boldsymbol{x}) = F\left(\frac{\beta_{0r} - \langle \boldsymbol{x}, \boldsymbol{\alpha} \rangle - \langle \boldsymbol{z}, \boldsymbol{\phi}_r \rangle}{\exp[\langle \boldsymbol{u}, \boldsymbol{\gamma} \rangle + \langle \boldsymbol{v}, \boldsymbol{\delta}_r \rangle]} \right) \quad (r = 1, \ldots, k-1) \qquad (3.23)$$

where $\boldsymbol{x}$ and $\boldsymbol{u}$ are the predictors, which are independent from the current category r, and the influence of $\boldsymbol{z}$ and $\boldsymbol{v}$ is specific to the category. We get the partial-proportional odds model with $\boldsymbol{u} = 0 = \boldsymbol{v}$ and a proportional-odds model with a proportional scale in the special case $\boldsymbol{x} = \boldsymbol{u}$ and $\boldsymbol{z} = 0 = \boldsymbol{v}$.

Finite Mixture Approach

A very flexible approach dealing with heterogeneity among groups of individuals is the finite mixture approach for cumulative models (Everitt, 1988; Everitt and Merette, 1990; Uebersax, 1999). Is is especially useful if we only have a latent underlying variable with individual perception. In this situation, we cannot assume that every observation has the same thresholds for the ordered output. Therefore, we have to assume C groups of observations having nearly equal thresholds and influence by modeling for group c:

$$P(Y_c \leq r|\boldsymbol{x}) = F(\beta_{0,c,r} - \langle \boldsymbol{x}, \boldsymbol{\beta}_c^- \rangle) \ , \qquad (3.24)$$

i.e. a separate proportional model for every group c.

However, individual class membership is not observable and we have to assume that each individual belongs to class c with the probability π_c. Thus, the complete model can be written as a finite mixture of the group probabilities (3.24)

$$P(Y \leq r|\boldsymbol{x}) = \sum_{c=1}^{C} \pi_c P(Y_c|\boldsymbol{x}; \boldsymbol{\beta}_c) \ .$$

Using an EM algorithm, the parameters can be estimated if the model is identifiable (Uebersax, 1999).

This model may also be used for testing the proportional odds assumption by stating the hypothesis $\pi_i = \cdots = \pi_C$ and $\boldsymbol{\beta}_1 = \cdots = \boldsymbol{\beta}_C$ by using a Likelihood Ratio (LR) test (Boes and Winkelmann, 2006).

3.3.2 Continuation-Ratio or Sequential Model

The continuation-ratio model by Cox (1988) is well-suited for sequential data. It has a strong relationship to the standard hazard models as it is their pendant for a categorized output variables (Greenland, 1994). It has been published by other authors under different names, i.e. hierarchical model by McCullagh and Nelder (1989, p. 160-164) and sequential model (Tutz, 1990, 1991).

Models are build by assuming $k - 1$ latent variables $U_r = f(\boldsymbol{x}) + \epsilon_r$ for the transfer from class r to class $r + 1$ with cut points θ_r. Now Y is determined sequentially: If $U_1 \leq \theta_1$, we get $Y = 1$ and otherwise $Y > 1$. In the first case the algorithm stops. Otherwise we have to calculate U_2 and compare it to θ_2 to decide if $Y = 2$ or $Y > 2$ and so on. In general, we have in step r

$$Y = r|Y \geq r \text{ if } U_r \leq \theta_r$$
$$\text{resp. } Y > r|Y \geq r \text{ if } U_r > \theta_r \ .$$

Now we can show that

$$P(Y = r|Y \geq r, \boldsymbol{x}) = F(\theta_r + f(\boldsymbol{x})) \tag{3.25}$$

with

$$P(Y = r|Y \geq r, \boldsymbol{x}) = P(U_r \leq \theta_r) = P(\epsilon_r \leq \theta_r + f(\boldsymbol{x})) = F(\theta_r + f(\boldsymbol{x}))$$

and additionally

$$P(Y = r|\boldsymbol{x}) = F(\theta_r + f(\boldsymbol{x})) \prod_{i=1}^{r-1} (1 - F(\theta_i + f(\boldsymbol{x}))) \ .$$

The model (3.25) can also be treated as a discrete version of the proportional hazards model (Cox, 1988; Greenland, 1994).

Cox (1988) distinguishes the forward from the backward continuation-ratio model. The forward model (3.25) is more popular than the other one, which is:

$$P(Y < r + 1|Y \leq r + 1; \boldsymbol{x}) = F(\theta_r + f(\boldsymbol{x}))$$

Continuation-ratio models are not reversible, i.e. reversing the order of the classes would lead to different estimators. This feature is recommended for models which describe a changing process over time (Greenland, 1994). Similar to the cumulative model, it is not invariant to output-stratified sampling (Greenland, 1994). Nevertheless, the continuation-ratio models are stochastically ordered in their proportional variant (Barnhart and Sampson, 1994).

Estimation
Using the logistic link function, these models can easily be estimated with the help of standard software by rearranging the data (Armstrong and Sloan, 1989). For instance[54], the model (3.25) in its linear form

$$P(Y = r|Y \geq r, \boldsymbol{x}) = F(\theta_r + \langle \boldsymbol{\beta}, \boldsymbol{x} \rangle)$$

with parameter vector $\boldsymbol{\beta}$, can be split up into a sequence of binary models for $r = 1$ up to $r = k - 1$. It is necessary to encode a binary variable y' and recode each line n from the design matrix $(1, x_{n1}, x_{n2}, \ldots, x_{nd})$ with corresponding $y_n = r$ to r lines with

y'	θ_0	θ_1	$\ldots$	θ_r	$\ldots$	θ_{k-1}	x_1	$\ldots$	x_d
0	1	0		0		0	x_{n1}		x_{nd}
0	0	1		0		0	x_{n1}		x_{nd}
$\vdots$						$\vdots$		$\vdots$	
1	0	0		1		0	x_{n1}	$\ldots$	x_{nd}

---

[54] The backward continuation-ratio model may be treated in an analogous way.

If $y_n = k$, we get $y' = 0$ for each of the $k - 1$ lines.

Besides this binary modeling, there is also a multivariate solution. Reformulate (3.25) to

$$P(Y = r|\boldsymbol{x}) = F(\theta_r + \langle\boldsymbol{\beta}, \boldsymbol{x}\rangle) \prod_{i=1}^{r-1} (1 - F(\theta_i + \langle\boldsymbol{\beta}, \boldsymbol{x}\rangle))$$

and use the response function

$$h = (h_1, \ldots, h_{k-1}) : \mathbb{R}^{k-1} \to \mathbb{R}^{k-1}$$

with the components

$$h_r(\eta_{i1}, \ldots, \eta_{i,k-1}) = F(\eta_i r) \prod_{j=1}^{r-1} (1 - F(\eta_i j)) \; .$$

Then, we can use the common techniques for unordered generalized linear models (cf. subsection 2.3.3) (Tutz, 1990). The link function is derived from (3.25) to

$$g = (g_1, \ldots, g_{k-1}) = \mathbb{R}^{k-1} \to \mathbb{R}$$

with components

$$g_r(\pi_1, \ldots, \pi_{k-1}) = F^{-1}\left(\frac{\pi_r}{1 - \pi_1 - \cdots - \pi_r}\right) \; .$$

Both modelings can be used in MLP networks for sequential ordinal output data, but in section 4.6 we will see, that the binary modeling has got some advantages.

3.3.3 Choice of a Model for Ordinal Data With a Continuous Underlying Variable

The two introduced models can be distinguished with the help of their model assumptions. While the cumulative odds model is invariant to reversing the categories, the continuation-ratio model is not (Armstrong and Sloan, 1989). Even, if the problem does not allow reversal of the categories, the proportional odds assumption could be more appropriate than the general continuation-

ratio model for falling conditional log odds ratios which decrease to zero
the less parameters are used (Armstrong and Sloan, 1989). Using the log-
log link in the continuation-ratio model with proportionality is equivalent to
the proportional cumulative model with the same link function (Läärä and
Matthews, 1985). A more general result is obtained by Tutz (1991). He shows
that the cumulative and the sequential models with linear functionals are
equivalent, if F is asymmetric. Furthermore, equivalence is also given, if F
is a solution of the functional equation $(1 - F(x))^2 = 1 - F(\alpha x + \beta x)$ with
arbitrary α and β (Tutz, 1991, p. 289).

Models with category independent parameter vectors $\boldsymbol{\beta}^-$ provide non-crossing decision boundaries. By incorporating the order restrictions $\beta_{0,1} < \cdots <
\beta_{0,k-1}$ we can state, that these models provide the two properties for decision
boundaries mentioned in 3.2. In respect to non-proportional models it can be
argued that the crossings occur outside of the interesting part of the $\mathbb{R}^d$; but
there is no mechanism which could ensure this.

The proportional models provide a stochastic ordering on the output distributions (Barnhart and Sampson, 1994). Additionally, the cumulative model
is reversible but the sequential is not. It is not sure, if the the output distribution is unimodal like it is required because there are no restrictions on
the probabilities. Regarding good-natured problems, it is given with high
probability.

The cumulative and the continuation-ratio models are inappropriate for the
analysis of output-stratified random sampling data because they are dependent
on the proportions of the output categories.

3.4 Models for Simple Ordinal Data

We have already discussed the approach of using an appropriate (non-)parame-
tric mapping function to deal with ordinal data (cf. subsection 3.1.1). In the
remainder of this section various parametric approaches for ordinal output data
modelings which are not based on the assumption of an underlying continuous
variable are described.

3.4.1 Adjacent-Category Model

The adjacent-category model (Simon, 1974) has been applied by Goodman (1983) and got its name by Agresti (1984) who described it in more detail (Agresti, 1999). This model and the continuation-ratio model are the most used and implemented methods of modeling ordinal response variables in linear regression (Bender and Benner, 2000; Lall, Campbell, Walters and Morgan, 2002). The adjacent-category model is given by

$$P(Y = r | Y \in \{r, r+1\}; \boldsymbol{x}) = F(\kappa_r - f(\boldsymbol{x})) \quad (s = 1, \ldots, k-1), \quad (3.26)$$

and its most common version is the linear logit

$$\text{logit}\left[P(Y = r | Y \in \{r, r+1\}; \boldsymbol{x})\right] = \beta_{0,r} - \langle \boldsymbol{\beta_r^-}, \boldsymbol{x} \rangle \quad (r = 1, \ldots, k-1) \ . \quad (3.27)$$

This model implies linear decision boundaries in the feature space, which might cross each other unless β_r was equal for every category r. This model implies a stochastic ordering of the response distributions for different predictor values (Barnhart and Sampson, 1994; Agresti, 1999) and it is reversible.

The probabilities for this model can be rewritten with the help of

$$\log\left[P(Y = r | Y \in \{r, r+1\}; \boldsymbol{x})\right] = \log\left[\frac{P(Y = r | \boldsymbol{x})}{P(Y = k | \boldsymbol{x})}\right] - \log\left[\frac{P(Y = r+1 | \boldsymbol{x})}{P(Y = k | \boldsymbol{x})}\right]$$

to

$$\log \frac{P(Y = r | \boldsymbol{x})}{P(Y = k | \boldsymbol{x})} = \sum_{s=r}^{k-1} \beta_{0,r} - \boldsymbol{\beta_r^{-t}}(k - s)\boldsymbol{x} \quad (3.28)$$

$$= \beta_{0,r}^* + \langle \boldsymbol{\beta_r^-}, \boldsymbol{u_r} \rangle$$

with $\boldsymbol{u_r} = (k - r)\boldsymbol{x}$. Therefore, standard algorithms for estimation in multinomial logistic regression can be used (Agresti, 1999, p. 286).

Equation (3.28) is a special version of

$$P(Y = r | \boldsymbol{x}) = \frac{\exp(\beta_r + \langle \boldsymbol{x}, \boldsymbol{\beta_j} \rangle)}{\sum_{s=1}^{k-1} \beta_s + \langle \boldsymbol{x}, \boldsymbol{\beta_s} \rangle}$$

which is invariant to output-stratified sampling (Greenland, 1994, p. 1670f.).

Obviously this model is related to Wang's modeling, which is an "inverted" variant of the adjacent-category model (Wang, 1986):

$$\log\left[P(Y = r + 1|\boldsymbol{x}, Y \in \{r, r+1\})\right] = \beta_{0,r} - \langle \boldsymbol{\beta}_{\boldsymbol{r}}^-, \boldsymbol{x} \rangle \quad (r = 1, \ldots, k - 1),$$

3.4.2 Stereotype Model

The linear stereotype model has been proposed by Anderson (1984). Its general form is

$$P(Y = r|\boldsymbol{x}) = F(\beta_{0,r} + \langle \boldsymbol{x}, \boldsymbol{\beta}^- s_r \rangle) \quad (r = 1, \ldots, k - 1) , \tag{3.29}$$

where $\beta_{0,0} = 0 = s_0$ and s_r denotes a scalar, which is interpreted as the score attached to category r. Constraints $s_1 \leq \ldots \leq s_k$ have been imposed in the original paper (Anderson, 1984, p. 6) and are supposed to be inessential unless subject-matter considerations indicate the contrary (Greenland, 1994, p. 1669). Therefore, this is a nonparametric scoring model at the same time.

The model is invariant under reversal of the categories, but produces changed estimators for $\boldsymbol{\beta}$ and $\boldsymbol{s}$ (Greenland, 1994, p. 1670). The decision boundaries are parallel and do not cross because their orientation is determined by the same $\boldsymbol{\beta}^-$ for all categories. Restrictions on the $\beta_{0,r}$ guarantee the ordering of the decision boundaries. Using the monotonic restrictions on the s_r the output distribution will be stochastically ordered.

The invariance of the model to output-stratified samples is a very interesting property (Greenland, 1994, p. 1670).

3.4.3 Choice of a Model for Ordinal Data Without a Continuous Underlying Variable

The adjacent category and the stereotype model are very similar. The probabilities are modeled directly for the stereotype model. In contrast, the adjacent category odds are modeled in the adjacent-categories model. Therefore the choice of the model depends on the intended interpretation of the estimated model parameter. For instance, if the adjacent odds are to be modeled, the adjacent-categories model should be used, and if the probabilities should be modeled directly, the stereotype model is appropriate. Using nonlinear functions instead of linear ones, both models will not provide essentially different estimates.

Although we can select models with and without underlying variable by simply looking at the type of the ordinal data, the choice is not unique. In some cases, no perfect model exists. Often, we have to choose between two models reflecting only some of the semantically possible assumptions. We have to decide, which assumption provides stronger restrictions on the set of possible models in order to retain as much information as possible in these cases.

3.5 Model Evaluations and Summary

The goodness-of-fit statistics are used for the evaluation of these models. Furthermore, residuals can be analyzed to detect model misspecifications.

3.5.1 Approaches for Ordinal Model Evaluation

The goodness-of-fit statistics for continuous explanatory variables are useless because they do not approximate the χ^2 distribution[55] (cf. subsection 2.3.3). Therefore, there are suggestions for alternative test statistics. First the statistics could be corrected to fit better with "small" samples or if—n_i representing the quantity of category i in a sample of size n—$\frac{n_i}{n} \to 0$ for some i respectively. The first approach is based on the calculation of exact distributions what is very expensive and will only be applicable for very small n. There are alternatives proposed based on scoring the output for the case $\frac{n_i}{n} \to 0$ (Lipsitz, Fitzmaurice and Molenberghs, 1996; Pulkstenis and Robinson, 2004). These approaches are sensitive to the chosen scoring systems which is usually the simple numbering $(1, \ldots, k)$ of the categories (Pulkstenis and Robinson, 2004).

In contrast to these approaches, which are based on ideas for categorical data, goodness-of-fit statistics for standard regression could be adapted. Agresti (1986) presents a statistic, which is based on the general proportional reduction in dispersion[56]

$$\frac{\sum_{i=1}^{n} D(Y_i) - \sum_{i=1}^{n} D(Y_i|\boldsymbol{X}_i)}{\sum_{i=1}^{n} D(Y_i)} \, , \tag{3.30}$$

---

[55]Nevertheless, differences of values of the deviance (2.15) may be used for model comparisons (Agresti and Yang, 1987)

[56]Measures based on this formula are of the r^2-type.

where $D(Y_i)$ describes the dispersion for the ith observation calculated relative to the estimated marginal distribution $(\hat{\pi}_1, \ldots, \hat{\pi}_k)$ and $D(Y_i|X_i)$ the measure computed for the estimated distribution $(\hat{\pi}_{1,i}, \ldots, \hat{\pi}_{k,i})$ conditional to the covariates x_i. Let $v_1 < v_2 < \cdots < v_k$ be scores for the categories, let $\hat{\mu} = \sum_s v_s \hat{\pi}_s$ and let $\hat{\mu}_i = \sum_s v_s \hat{\pi}_{s,i}$. Now let Y_i denote the associated score for observation i, i.e. $Y_i = v_s$ if the ith response is in the sth category. Then the proportional reduction in variance measure is obtained:

$$\hat{\eta} = \frac{\sum_i (Y_i - \hat{\mu})^2 - \sum_i (Y_i - \hat{\mu}_i)^2}{\sum_i (Y_i - \hat{\mu})^2} \tag{3.31}$$

This is also based on scores associated to the ordinal data. Therefore, it is suited for data with continuous underlying. This can be done by choosing scores, which represent the average categories, e.g. the midpoints between the thresholds. The general form (3.30) should be used for simple ordinal data with the help of an ordinal measure of dispersion.

Further, we might be interested in testing the goodness-of-link. However— especially in the remainder of this thesis—nonlinear models do *not* depend on the choice of the link function (Cox, 1984) and that is the reason why these tests are not covered in this thesis.[57]

3.5.2 Example: Lecture Evaluation

This example illustrates the results from this chapter, especially the practical usefulness of the symmetry assumption is shown. The bigger part of the calculations has been done with SPSS® for Windows® version 15.[58]

Description of the Sample
Students at the University of Münster are asked to take part in an electronic survey for evaluation of courses. The questionnaire is written in German and printed in appendix B. It is only slightly altered for the different courses. Here we have 383 participants of the course „Mathematik für Wirtschaftswissenschaftler" (mathematics for economic analysis), most of them studying business administration (236), information systems (71) and economics (57). 211 of them are male, 136 are female and 36 are missing. 203 have booked basic

---

[57]cf. Genter and Farewell (1985) for instance.

[58]There are solutions for SAS® and S-Plus®, too (Bender and Benner, 2000).

courses („Grundkurs") in mathematics during their senior years of secondary school („Gymnasiale Oberstufe"), 141 of them have also been examined in mathematics during their final examination („Abitur"). 169 took advanced courses („Leistungskurs") in mathematics at school. 324 students have missed less than 10% of the lessons, 37 missed $11 - 20\%$ and only 9 did not visit more than 20% of the lessons.

Some of the questions allow abstention, what is handled like missing values to make analysis easier.

Methods
The aim of this analysis is to identify the explanatory variables for the students' satisfaction with the course (question 7.6, variable number 24). Therefore the following questions have been included as explanatory variables into the model: 2.1 to 7.2, 7.5, 9.2 (cf. appendix B).

Three different analyses have been made in order to compare and discuss the results:

1. direct regression (dir. regression)

2. regression with standardized variables (std. regression)
   All variables have been "standardized" on the interval $[0, 1]$, e.g. for five categories the values have been transformed to $\{0, 0.25, 0.5, 0.75, 1\}$.

3. proportional odds logistic regression with symmetric inputs
   We have assumed, that variables 5 to 20 and 23 are symmetric in the perception of the participant (cf. subsection 3.1.2).

Results
The goodness-of-fit values for all three models lie in the same range (at about 0.5) indicating that only about half of the variation could be explained by the included variables. The detailed results are shown in Table 3.2. The big differences between the regression models on the original and the rescaled data indicate, that the functional relationship is not linear.

The variables 5, 6, 9, 12, 17 and 23 are significant in all three models (on the 10% level). The variables 11 and 15 are only significant in the first and the third model. The variables 10 and 20 are only significant in the ordinal regression model.

			model	
quest.var.		dir. regression	std. regression	ord. regression
2.1	5	.145 (.010)	.156 (.012)	1.480 (.036)
2.2	6	-.143 (.010)	-.149 (.015)	-1.509 (.031)
2.3	7	-.007 (.923)	.018 (.826)	.009 (.753)
3.1	8	-.002 (.976)	-.037 (.594)	.475 (.414)
3.2	9	.148 (.023)	.250 (.001)	2.038 (.009)
3.3	10	-.063 (.245)	-.002 (.978)	-1.010 (.100)
4.1	11	.132 (.007)	.083 (.167)	2.079 (.003)
4.2	12	.168 (.002)	.159 (.017)	2.173 (.008)
4.3	13	.030 (.603)	.025 (.721)	0.996 (.267)
5.1	14	-.003 (.975)	.016 (.787)	-.256 (.688)
5.2	15	.091 (.095)	.044 (.471)	1.007 (.042)
5.3	16	.024 (.600)	.015 (.780)	0.344 (.626)
6.1	17	.146 (.000)	.150 (.002)	1.600 (.006)
6.2	18	.228 (.000)	.149 (.041)	2.356 (.001)
7.1	19	.044 (.264)	.056 (.244)	.557 (.172)
7.2	20	.048 (.177)	.017 (.697)	.716 (.036)
7.5	23	-.226 (.000)	-.152 (.091)	-3.157 (.000)
9.2	29	.099 (.027)	-.020 (.778)	.000 (.427)
R^2		.559	.507	.380–.649

Tab. 3.2: Explanatory variables and their values $\hat{beta}_i$ (significance)

The interesting question which has to be answered with this kind of analysis is: How could the lecturer improve his course in order to increase the students' satisfaction?

First, the satisfaction is not dependent on the students' prior knowledge on mathematics. Therefore, the course is arranged in an appropriate way. The presentation of the content is strongly correlated with the students' satisfaction. But as the students' satisfaction with these characteristics is quite well[59], it is difficult to improve the course at these aspects. Questions 3.2 and 4.1, which represent motivation and expressive capability of the lecturer are also

---

[59](11.8%, 44.4%, 25.1%, 16.3%, 2.5%) and (19.7%, 57.5%, 21.1%, 1.4%, 0.3%) from very content to very discontent at questions 6.1 and 6.2, respectively

important and high values of satisfaction[60]. However, the students have problems in understanding the contents of the course which is indicated by relatively bad values[61] for question 4.2 which also has a strong influence on the students' satisfaction with the course. It is very plausible, that this is the main influence on the students' satisfaction, which is also supported by the strong inverse influence of question 7.5.

Surprisingly, the two ordinary regression models produce comparable results (regarding the significance of the variables) to the ordinal model. However, the ordinal regression model does not need any additional assumptions on the scaling and the error variable. Therefore, it is the more reliable model. The importance of possible changes in scaling can be seen from comparing variable significance in direct and standardized regression. The variables 11, 15, 29 and even 20 are significant in direct regression in contrast to standardized regression.

This small example vividly demonstrates, that the ordinal regression model with ordinal encoded explanatory variables is well qualified to replace the standard linear regression for ordinal variables. And there is enough information provided to analyze and interpret surveys, as the estimated parameters indicate the direction of and sensitivity to the explanatory variables' influence.

3.5.3 Implications for Nonlinear Ordinal Regression

The introduced parametric approach to ordinal input data modeling can be applied to the framework of GLM by recoding the Design matrix appropriately and incorporating constraints on the parameters, i.e. for estimation a constraint variant of the IRLS approach is required and may be found in general techniques for constraint nonlinear optimization algorithms (Agresti, Chuang and Kezouh, 1987). This is also given for ordinal output data models, where we may use one of the different link functions with associated constraints (see Table 3.3).

None of the introduced models ensures a unimodal output distribution. In general, this is automatically given in proportional models with a fixed linear parameter β^- for all categories. Obviously, this is not easy for nonproportional decision boundaries.

Extensions to semiparametric models for ordinal data have been proposed with the help of additional smoothing components to the linear predictor

---

[60](12.1%, 48.3%, 33.0%, 6.4%, 0.3%) and (17.6%, 44.4%, 29.5%, 7.9%, 0.5%)
[61](5.9%, 34.1%, 36.8%, 17.7%, 5.4%)

model	link	reparametrization	constraints
cumulative	$F^{-1}(\pi_1 + \cdots + \pi_r)$	$\theta_1^* = \theta_1$ $\theta_r^* = \log[\theta_r - \theta_{r-1}]$	none
continuation-ratio	$F^{-1}\left(\frac{\pi_r}{1 - \pi_1 - \cdots - \pi_{r-1}}\right)$	none	$\theta_1 < \cdots < \theta_{k-1}$
adjacent-categories	$F^{-1}\left(\frac{\pi_r}{\pi_k}\right)$	$\beta_{0,r}^* = \sum_{s=r}^{k-1} \beta_{0,r}$ $x_r^* = (k-r)x$	none
stereotype	$F^{-1}(\pi_r)$	none	$\theta_1 < \cdots < \theta_{k-1},$ scaled proportionality

Tab. 3.3: Variants of link functions for ordinal regression

(Tutz, 2003) and generalization of the used link function (Stewart, 2005). The approach of Tutz (2003) supports nonlinear decision boundaries via additive smoothing components. We would like to use MLP networks for nonlinear analysis of ordinal and metric data, which are presented in the next chapter.

4 Multi-layer Perceptron Networks

From a statistical point of view Multi-layer Perceptron (MLP) networks are the best analyzed neural network techniques in literature.[62] The term neural network is not exactly specified and the area of neural networks covers a lot of algorithm-driven techniques[63]. This is the reason, why a common background for these models is hard to identify (cf. section 2.5).

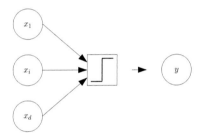

Fig. 4.1: Perceptron

The term MLP has grown historically, and it describes multi-layered feed-forward networks in general. The term feed-forward describes a type of neural networks, in which the flow of calculation is directed and any recurrence is not allowed. The perceptron is a neuron (node), which is build up as illustrated in Figure 4.1 and represents a function $\mathbb{R}^d \to \mathbb{R}$ with $w_j \in \mathbb{R}$ for all j:

$$y = g\Big(\sum_{j=0}^{d} w_j x_j\Big) \quad \text{with } g(a) = \begin{cases} -1 & a < 0 \\ 1 & a \geq 0 \end{cases}$$

The transfer function $g(\cdot)$ has been this threshold function in its original

[62]See Amari (1990), Sarle (1994), Cheng and Titterington (1994), Warner and Misra (1996), Ripley (1993, 1994b, 1997), Ripley and Mardia (1994), White (1989b) for some general aspects and Schumacher, Roßner and Vach (1996); Vach, Roßner and Schumacher (1996) for aspects related to logistic regression.

[63]See Reed and Marks (1998) for algorithmic contributions to MLP networks.

definition (Rosenblatt, 1958, 1962). Nevertheless, it has been generalized very early to sigmoidal[64] functions building a relation to logistic regression[65]. Thus a multi-layer feed-forward network is made up of generalized perceptrons and the term MLP is widely accepted (Bishop, 1995). Another term that is often used for this type of network is Error Back-propagation Network (EBPN), which is directly related to the characteristic optimization algorithm. Feed-forward Neural Network (FFNN) is another term often used for this type of neural network (Fine, 1999), but it describes a whole class of neural networks containing the perceptron, for example the MLP and the Adaptive Linear Neuron (ADALINE) (Widrow and Hoff, 1960).

Because ordinal data analysis with MLP networks is one of the main topics of this thesis, this chapter deals with the major properties of MLP networks in section 4.1 and 4.4 as well as techniques for estimation in section 4.2 and 4.3. Classification with MLP networks is discussed in section 4.5 before ordinal regression with MLP networks is described in 4.6.

4.1 Model and its Expressive Power

The most usual[66] MLP network consist of a single hidden layer (a layer that is not directly related to input or output, i.e. its values are unobserved, see Figure 4.2) and represents the function (Ripley, 1996, p. 143)

$$f : \mathbb{R}^d \to \mathbb{R}^k$$

$$f_r(\boldsymbol{X}|\boldsymbol{w}) = g_r^{(2)} \left(w_r^{(2)} + \sum_{j \to r} w_{jr}^{(2)} g_j^{(1)} \left(w_j^{(1)} + \sum_{i \to j} w_{ij}^{(1)} x_i \right) \right) \qquad (4.1)$$

with $w_{ij}^{(k)}, w_i^{(k)} \in \mathbb{R}$ for all i, j, k. The functions $g_j^{(1)} : \mathbb{R} \to \mathbb{R}$ are usually identical for all j (sigmoidal) and independent from the type of analysis the network is used for. The functions $g_r^{(2)} : \mathbb{R} \to \mathbb{R}$ are also chosen identical, but they are dependent on the type of analysis the network has been designed for, i.e. classification ore regression. In case of regression, it is usually the identity while especially for classification it is also of sigmoidal type. The choice for $g_r^{(2)}$

---

[64]Differentiable, isotone and bounded function $\mathbb{R} \to \mathbb{R}$ (see subsection 4.1.1).
[65]More historical remarks on neural networks may be found in Haykin (1999).
[66]see (Bishop, 1995, p. 130).

is strongly related to the choice made for the response functions[67] in GLMs (McCullagh and Nelder, 1989, p. 30f.).

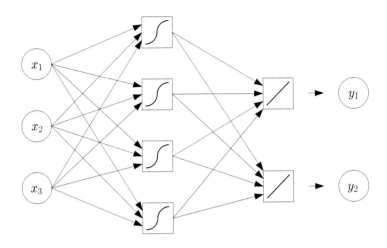

Fig. 4.2: MLP network example

The single perceptron is well-suited to estimate a linear decision boundary and thus is the "neural network version" of the LDA. By contrast, the MLP is capable to estimate nonlinear decision boundaries as well (Bishop, 1995; Haykin, 1999; Fine, 1999).

4.1.1 Representational Capability of Multi-layer Perceptron Networks

The space of all representationable functions has to be known to analyze the representational capabilities of MLP networks. Therefore we define:

Definition 4.1 (set of one hidden layer MLP networks)
The class of functions

$$\mathfrak{N}_{d,k}^{(h)}(g) = \left\{ f : \mathbb{R}^d \to \mathbb{R}^k : f_k(x) = \sum_{j=1}^{h} w_j^{(2)} g\left(\langle w_j^{(1)}, x \rangle + w_0^{(1)} \right) \right\} ,$$

---

[67]inverse of the link function

where $g : \mathbb{R} - \mathbb{R}$ *denotes the* transfer function *and* $h \in \mathbb{N}$ *is the number of hidden nodes, is called the* set of one hidden layer MLP networks.

Comment 4.2

For example, the neural network in Figure 4.2 is a network from class $\mathfrak{N}_{3,2}^{(4)}(g)$ *with a sigmoidal, but yet not exactly specified function g. Furthermore, the network function in this definition is related to equation* (4.1) *by setting the transfer functions at the output nodes to the identity.*

In the next paragraphs we will characterize $\mathfrak{N}_{d,k}^{(h)}(g)$ for certain transfer functions g and investigate, if all integrable functions can be represented by a MLP, that has a sufficient size and is made up of appropriate transfer functions.

The term "sigmoid function" describes the most important type of transfer functions g and is defined as follows:

Definition 4.3 (sigmoid function)
A function $t : \mathbb{R} \to \mathbb{R}$ *is called a* sigmoid function, *if and only if it is bounded and monotonically increasing, i.e.*

$$\lim_{x \to -\infty} t(x) > -\infty \text{ and } \lim_{x \to \infty} t(x) < \infty \text{ and } x < y \Rightarrow g(x) \leq g(y) \ .$$

Comment 4.4

Often, especially for the reason of easier calculations, the transfer function should be differentiable, i.e. depending on the type of optimization algorithm $g \in \mathcal{C}^{\geq 1}(\mathbb{R})$ *for first-order or* $g \in \mathcal{C}^{\geq 2}(\mathbb{R})$ *for second-order methods (Bishop, 1995; Reed and Marks, 1998; Fine, 1999). In these cases, the sigmoidal function has the typical s-shape, which motivated the name.*

Exact Implementation via Kolmogorov's Solution
Beginning with Kolmogorov's solution to Hilbert's 13th problem, substantial work has been done to obtain the following result (Sprecher, 1965a,b, 1966):

Theorem 4.5 (Sprecher, 1993)
Let λ_k *be a sequence of positive integrally independent numbers; i.e.* $\sum_k r_k \lambda_k \neq 0$ *for any integers* r_k *that are not all 0. Letting* $D = [0, 1 + \frac{1}{5!}]$, *there exists a continuous monotonically increasing function* $\psi : D \to D$ *having the following property:*

For every real-valued continuous function $f : [0,1]^d \rightarrow \mathbb{R}$ with $d \geq 2$ there are continuous functions $\Phi_k : \mathbb{R} \rightarrow \mathbb{R}$ and constants $\{a_d\}$, $\{b_m\}$ such that

$$f(x_1, \ldots, x_d) = \sum_{k=0}^{2d} \Phi_k \left[b_k + \sum_{j=1}^{d} \lambda_j \psi(x_j + ka_d) \right] .$$

This result has been made applicable to MLP networks by Hecht-Nielsen (1987). Kůrková (1992) extends this result by providing estimations for the minimal number of hidden nodes that is sufficient for representing certain functions:[68]

Theorem 4.6 (Kůrková, 1992, p. 503)

Let $n \in \mathbb{N}$ with $n \geq 2$, $g : \mathbb{R} \rightarrow [0,1]$ be a sigmoidal function, $f^* \in \mathcal{C}([0,1])$ and $0 < \epsilon \in \mathbb{R}$. Then for every $m \in \mathbb{N}$ such that $m \geq 2n+1$ and $\frac{n}{(m-n)} + v < \frac{\epsilon}{\|f^*\|}$ and $\omega_{f^*}(\frac{1}{m}) < \frac{v(m-n)}{2m-3n}$ for some positive real v, f^* can be approximated with an accuracy ϵ by a MLP with two hidden layers, containing $nm(m+1)$ nodes in the first hidden layer and $m^2(m+1)^n$ nodes in the second layer, with a transfer function g in such a way that all weights and biases, with the exception of weights corresponding to the transfer from the second hidden layer to the output node, are universal for all functions f with $\|f\| \leq \|f^*\|$ and $\omega_g \leq \omega_f$.[69]

The aim of these approaches is to represent functions exactly. That is the reason why they often need an arbitrarily large number of nodes in two hidden layers, so that these results are of limited use with respect to statistical regression.

Function Approximation for Continuous Functions

One of the first results to characterize the space of $\mathfrak{N}_{d,1}(g)$ needs the following definition (Cybenco, 1989, p. 306):

---

[68] $\omega_f(\delta) = \sup\{\|f(x_1, \ldots, x_d) - f(y_1, \ldots, y_d)\|, (x_1, \ldots, x_d), (y_1, \ldots, y_d) \in [0,1]^d : |x_p - y_p| < \delta \quad \forall p = 1, \ldots, d\}$

[69] The weights and biases are universal in that sense, that they are independent from a concrete f.

Definition 4.7 (discriminatory function)
We say that g is discriminatory *if for any finite, signed, regular measure[70] μ*

$$\int_{[0,1]^d} g(y^t x + \theta)\, d\mu(x) = 0$$

for all $y \in \mathbb{R}^d$ and $\theta \in \mathbb{R}$ implies $\mu = 0$.

Then, we can state:

Theorem 4.8 (Cybenco, 1989, p. 306)
If the transfer function g is continuous and discriminatory, then $\mathfrak{N}_{d,1}(g)$ is dense in $\mathcal{C}([0,1]^d)$.

This result may be reformulated by using a more restrictive assumption on g and relaxing the represented class:

Theorem 4.9 (Hornik, Stinchcombe and White, 1989)
For $d \in \mathbb{N}$, $g : \mathbb{R} \to [0,1]$ is any CDF. Then $\mathfrak{N}_{d,1}(g)$ is uniformly dense on compacta in $\mathcal{C}(\mathbb{R}^d)$.

This is sufficient for solving practical regression problems with MLP networks (Kuan and White, 1994). Nevertheless, it is possible to relax the assumptions on g and to extend the class of transfer functions. First, let us have a look at the relaxed assumptions on g.

Comment 4.10
Stinchcombe and White (1989) extend this result to nonlinear continuous $g \in L^1(\mathbb{R}, \lambda)$ and Funahashi (1989) to nonconstant, bounded and monotone increasing g.

Comment 4.11
This result empowers us to state, that MLP networks are universal function approximators. In case of classification, the output of the neural network is often restricted to be in $[0,1]$. This is usually achieved by using a squashing function, i.e. a function $\mathbb{R} \to (a,b)$ with $a, b \in \mathbb{R}$ and $b - a < \infty$. Castro, Mantas and Benítez (2000) prove, that using an arbitrary squashing function

---

[70]This is a very large class of measures, which especially includes the Borel measures on $\mathbb{R}^n$, $n \in \mathbb{N}$. More detailed discussion can be found in Cohn (1980).

$\mathbb{R} \rightarrow [0, 1]$ *within the output layer of a MLP can be used to approximate arbitrary density distributions.*

Further results make use of the following definition for relaxations of the terms "continuous" and "bounded":

Definition 4.12
Let $\mathcal{M}$ denote the set of node functions g such that

1. *The closure of the set of points of discontinuity of any $g \in \mathcal{M}$ has zero Lebesgue measure.*

2. *For every compact set $K \subset \mathbb{R}$, the essential supremum[71] of g on K, with respect to Lebesgue measure λ, is bounded by*

$$\inf\{\lambda : \lambda\{x : x \in K, |g(x)| \geq \lambda\} = 0\} \ .$$

Theorem 4.13 (Leshno, Lin, Pinkus and Schocken, 1993, p. 863)
If $g \in \mathcal{M}$, then $\mathfrak{N}_{d,1}(g)$ is dense in $\mathcal{C}(\mathbb{R}^d)$ if and only if g is not almost everywhere an algebraic polynomial.

Function Approximation for Integrable Functions
Another way of relaxation is to allow a greater function space. For instance, the space of integrable functions:

Definition 4.14

$$\mathcal{L}_p(\mathfrak{X}) = \{f : \left[\int_{\mathfrak{X}} |f|^p \, d\mu\right]^{\frac{1}{p}} < \infty\}$$

Here, the first result is based on a slightly different assumption on g to obtain a result for $L_1(\mathfrak{X})$

Theorem 4.15 (Cybenco, 1989, p. 310)
If $\mathfrak{X} \in \mathbb{R}^d$ is compact and the node function g is measurable and sigmoidal, then $\mathfrak{N}_{d,1}(g)$ is dense in $\mathcal{L}^1(\mathfrak{X})$ with respect to the L^1 metric.

---

[71]supremum restricted to the subsets with positive measure

By the same time Hecht-Nielsen (1989) obtained similar results for $\mathcal{L}^2([0,1]^d)$ by using stronger assumptions on f but weaker ones on the function g:

Theorem 4.16 (Hecht-Nielsen, 1989, p. 597)
Given any $\epsilon > 0$ and any $\mathcal{L}^2([0,1]^d)$ function $f : [0,1]^d \subset \mathbb{R}^d \to \mathbb{R}^k$, there exists a MLP network with one hidden layer with sigmoidal transfer functions g that can approximate f to within ϵ mean squared error accuracy.

Depending on assumptions on the underlying measure μ two results have been published. First, Leshno, Lin, Pinkus and Schocken (1993) use a non-polynomial g and an absolutely continuous measure:

Theorem 4.17 (Leshno, Lin, Pinkus and Schocken, 1993, p. 863)
Given a non-negative measure μ that is absolutely continuous with respect to the Lebesgue measure. Let $g \in \mathcal{M}$. Then $\mathfrak{N}_{d,1}(g)$ is dense in $\mathcal{L}^p(\mu)$ in the sense of the L^p metric, so long as g is not almost everywhere a polynomial and $1 \le p < \infty$.

However, Hornik (1991) obtains a similar result by using a more relaxed g with a finite measure μ:

Theorem 4.18 (Hornik, 1991, p. 252)
If g is unbounded and non-constant, then $\mathfrak{N}_{d,1}(g)$ is dense in $L^p(\mu)$ for all finite measures μ on $\mathbb{R}^d$ and $1 \le p < \infty$.

These results (as well as those of the preceding paragraph) are presented in Table 4.1.

The results of this and the preceding paragraph are intended for use with regression. On basis of GLM, we can use regression to estimate the a-posteriori distributions. This is the reason, why we can transfer these results to classification. It is plausible to assume continuous functions for calculating the predictor η, because slight changes in the predictor values result in slight changes of the distribution function. This behavior is usually recommended for a-posteriori distributions in classification.

Therefore, MLP networks are well suited for classification. This can also be shown in a more direct way. Based on the work by Ruck, Rogers, Kabrisky, Oxley and Suter (1990), Funahashi (1998) has additionally shown, that a three-layered neural network with at least $2n$ neurons is able to approximate the

function type and transformations	function space
$\mathfrak{N}_{d,1}(g)$, g continuous, discriminatory	$\mathcal{C}([0,1]^d)$
$\mathfrak{N}_{d,1}(g)$, g CDF	$\mathcal{C}(\mathcal{X})$, $\mathcal{X} \subset \mathbb{R}^d$ compact
$\mathfrak{N}_{d,1}(g)$, g continuous, bounded and non-constant	$\mathcal{C}(\mathcal{X})$, $\mathcal{X} \subset \mathbb{R}^d$ compact
$\mathfrak{N}_{d,1}(g)$, $g \in L^1\mathbb{R}$	$\mathcal{L}^1([0,1]^d)$
$\mathfrak{N}_{d,1}(g)$, g bounded, measurable, sigmoidal	$\mathcal{L}^1(\mathcal{X})$, $\mathcal{X} \subset \mathbb{R}^d$ compact
$\mathfrak{N}_{d,1}(g)$, $w_0^{(1)} \in \mathbb{R}^d$, $g \in \mathcal{L}^1(\mathbb{R}^d)$	$\mathcal{C}(\mathbb{R}^d)$
$\mathfrak{N}_{d,1}(g)$, g sigmoidal	$\mathcal{L}^2([0,1])$
$\mathfrak{N}_{d,1}(g)$, $g \in \mathcal{M}$ not almost everywhere algebraic polynomial	$\mathcal{L}^p(\mu)$, $1 \leq p < \infty$
$\mathfrak{N}_{d,1}(g)$, g unbounded, non-constant	$\mathcal{L}^p(\mu)$, μ finite measure on $\mathbb{R}^d$

Tab. 4.1: Approximation Properties of MLP

a-posteriori probability in a two-category classification problem with arbitrary accuracy.

Approximation Performance

It is nice to know, that a MLP can approximate all feasible functions. Unfortunately, a very important detail has been left out: The number of nodes in the hidden layer, that is needed to obtain an approximation within some fixed bounds. Obviously, this number is influenced by various parameters like the choice of transfer functions (maybe different for every node), the number of hidden nodes and even by the performed data preprocessing.

There are two different problems considered in this section: First, we have to solve the problem of determining the number of hidden nodes to separate a set of vectors; i.e. discriminating two groups of real valued vectors. The second problem is to determine the speed of convergence for approximating a given function.

For the first case—which is similar to the VC dimension—the following theorem holds:

Theorem 4.19 (Baum, 1988, p. 199)
A one-hidden-layer MLP with $\lceil \frac{N}{d} \rceil$ hidden nodes with sigmoidal transfer functions can compute an arbitrary dichotomy on N d-dimensional vectors in general position.

Within the same work Baum (1988) proves, that

Theorem 4.20 (Baum, 1988, p. 205)
A one-hidden-layer MLP with $h = \lfloor \frac{4N}{d} \rfloor \lceil \frac{e}{\lfloor \log_2(\frac{N}{d}) \rfloor} \rceil$ hidden nodes is capable of computing an arbitrary mapping.

These results are obviously of little interest, because they only provide information on how many nodes are needed at maximum. If you would use the calculated numbers from Theorem 4.19 for classification and Theorem 4.20 for regression respectively, no generalization would take place. Therefore, these upper bounds are only of interest in calculating the VC-dimensions.[72]

For practical issues, the additional benefit, that could be obtained by adding nodes to the hidden layer, is of greater interest. In a quite general setting the following theorem holds:

Theorem 4.21 (Jones, 1992)
Let $f(\cdot)$ be a neural network like in (4.1) with h hidden nodes, $k = 1$, the transfer functions $g_j(\cdot)$ fixed and the $|w_j^{(2)}|$ are summable, then the rate of convergence is $O(\frac{1}{\sqrt{h}})$.

Therefore, adding nodes reduces the approximation error by a factor in $O(\frac{1}{\sqrt{h}})$ *independently* from the dimensionality of the feature space d. This rate of convergence has been calculated more precisely by Barron (1993, 1994); Hornik, Stinchcombe, White and Auer (1994). Lewicki and Marino (2004) have generalized it without improving it substantially.

Again, these results do not really help us in designing an appropriate MLP network.

4.1.2 Sigmoid Functions

The choice of the "correct" transfer function for MLP networks has been discussed for a long time in literature (cf. Brown, An, Harris and Wang (1993)).

---

[72]cf. (Vapnik, 1998) for detailed description of the VC theory.

Although every CDF will do the job for adaptable representation of any continuous functions (cf. Theorem 4.9), the relation between the used transfer function and the number of required nodes in the hidden layer to obey given approximation bounds is still unclear.

Mhaskar and Micchelli (1992) have shown that *any arbitrary* sigmoid function can be used in feed-forward neural networks for approximation of a given continuous function on a compact subset $k \in \mathbb{R}^n$ with a maximum error bounded by a fixed ϵ. Therefore, a lot of sigmoid functions have been proposed for use in feed-forward neural networks. The most common ones are (Bishop, 1995, p. 126f.):

- logistic function

$$t(a) := \frac{1}{1 + e^{-a}}$$

- hyperbolic tangent[73]

$$t(a) := \frac{1 + \tanh(a)}{2} \quad \text{with} \quad \tanh(a) := \frac{e^a - e^{-a}}{e^a + e^{-a}}$$

Both functions are solutions to the differential equation $\frac{df(x)}{dx} = f(x)(1 - f(x))$ and as a result, the derivatives, that are essential for gradient descent in backpropagation, can easily be calculated. This close relationship also provides the identity $(1 + e^{-x})^{-1} = \frac{1}{2}\left(1 + \tanh\left(\frac{1}{2}x\right)\right)$ (Anders, 1997, p. 48).

Mendil and Benmahammed (1999) proposed the transfer function

$$t(a) = \frac{a}{|a| + 1}$$

with derivative

$$t'(a) = (1 - |g(a)|)^2 \ ,$$

which can be calculated easier than the logistic function and the hyperbolic tangent, because the latter ones have to be approximated within digital computers.

---

[73]The hyperbolic tangent is often used without linear transformation to the domain $[0, 1]$ (Bishop, 1995).

There exists a long discussion on choosing the "right" sigmoidal transfer function.[74]

Practical results seem to support the hyperbolic tangent over the logistic function (Bishop, 1995, p. 127).

Even in the restricted case of sigmoidal transfer functions, the influence of the choice of the transfer function on the number of necessary nodes in the hidden layer is still under research. Mhaskar and Micchelli (1992) show for example, that with k-th degree sigmoid functions, i.e. functions having the properties $\lim_{x \to -\infty} \frac{f(x)}{x^k} = 0$, $\lim_{x \to \infty} \frac{f(x)}{x^k} = 1$ and $|f(x)| \leq c(1 + |x|)^k$ for some constant $c > 1$, for a fixed error in approximating a continuous function the number of required nodes decreases with k.

Generally speaking, the transfer function could be any CDF. This is why continuous distributions with an additional scale parameter are often proposed to control the slope of the transfer function (Reed and Marks, 1998, p. 132f.). Thimm, Fiesler and Moerland (1995) as well as Thimm, Moerland and Fiesler (1996) have shown, that this parameter is not identifiable within the MLP model and therefore is superfluous.

Generally, the use of CDFs of unimodal distribution functions is recommended. An arbitrary CDF may also belong to a multi-modal distribution having small "plateaus" within its increase. Obviously, these multi-modal distribution functions can be described as a finite mixture of unimodal distribution functions and thus are of a less general nature than the unimodal ones. This result can be transferred to neural network modeling: Nodes with multi-modal CDFs can be replaced by a certain number of nodes with unimodal transfer functions. For decision tasks, the (fixed) shape of the CDF is not so relevant, as it only reduces (in the multi-modal case) the increase in probability change where the decision boundary is crossed (see Figure 4.3).

4.1.3 Constructing Multi-layer Perceptron Networks

In addition to the results presented so far, there are some constructive approaches based on the Kolmogorov theorem (Jones, 1990; Nakamura, Mines and Kreinovich, 1993; Katsuura and Sprecher, 1994; Sprecher, 1996b,a) or based on ridge functions (Kůrková, 1992). But for our purposes, these results are

---

[74]cf. Bishop (1995, p. 127), Daqi and Genxing (2003), Mhaskar (1993); Mhaskar and Micchelli (1994), Fine (1999) and many others.

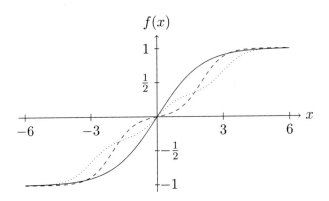

Fig. 4.3: Unimodal (line), bimodal (dashed) and trimodal (dotted) logistic
transfer function

irrelevant, because we do not know the target function exactly. The function
is only known at some distinct points and only measured with some error.

MLP networks suffer from the *Curse of Dimensionality*. The results from
subsection 4.1.1 are independent from the dimensionality of the input and
output space d and k respectively. Nevertheless, the number of nodes is left
open in these theorems, and it is directly affected by the dimensionality of the
input and output spaces (Fine, 1999). Mhaskar and Micchelli (1994, p. 69)
have shown, that $(d + 1)\epsilon^d + 2$ sigmoidal nodes are needed to obtain the
approximation up to an error of only ϵ.

Although these theoretical results give confidence that MLP networks are
suited for a wide range of regression and classification tasks, they shed little
light on how to design and estimate MLPs networks in practice (Duda, Hart
and Stork, 2000, p. 288). Therefore, we have to use methods of statistical
learning like bootstrapping, cross-validation, jackknifing and so on (Hastie,
Tibshirani and Friedman, 2001). We will come back to this topic in section 4.4.

4.2 Estimation with the Backpropagation Algorithm

The term *backpropagation* has been introduced by Rumelhart, Hinton and Williams (1986b,a), although the idea for calculating the derivatives in this stepwise manner has already been described much earlier.[75]

Bishop (1995) emphasizes the general nature of the backpropagation idea for calculating the derivatives of the error function. Therefore, backpropagation in the wide sense is an algorithm for estimating MLP networks. In the narrow sense it is an algorithm for calculating only the derivatives of the network function. In its wider sense, the backpropagation algorithm is usually used together with the gradient descent algorithm for calculating the weight updates during estimation. Nevertheless, we might use other optimization algorithms (cf. section 4.3) and therefore, we will refer to backpropagation as an algorithm for calculating the derivatives of the network function.

4.2.1 Backpropagation Algorithm

Bishop (1995, pp. 141–144) derives the backpropagation algorithm for general feed-forward networks having arbitrary differentiable nonlinear transfer functions and for arbitrary differentiable error function which can be written as a sum

$$E = \sum_n E^{(n)}$$

over the patterns in the training set. In the case of the MSE function this would be equation (2.5) with f as in equation (4.1).

Using the chain rule, the derivative of $E^{(n)}$ with respect to every w can be written as

$$\frac{\partial E^{(n)}}{\partial w_{ij}} = \delta_j z_i^{(n)}. \tag{4.2}$$

The basic equations for the different δ_j are

$$\delta_k = g_k'(a_k^{(n)}) \frac{\partial E^{(n)}}{\partial y_k} \tag{4.3}$$

---

[75] see Haykin (1999, p. 43) for some historical remarks.

for every output node k and

$$\delta_i = g_i'(a_i^{(n)}) \sum_k (w_{ik}\delta_k) \tag{4.4}$$

for every node i in the hidden layer.

The algorithm for evaluating the derivatives of the error $E^{(n)}$ with respect to the weights w can be summarized in four steps (Bishop, 1995, p. 144):[76]

Algorithm 4.1

1. *Evaluate the network (and thus obtain the values for all z_j) by applying the input vector $\boldsymbol{x}^{(n)}$.*

2. *Evaluate the δ_k for all output nodes using (4.3).*

3. *Back-propagate the δ's using (4.4) to obtain δ_j for each hidden node in the network.*

4. *Use (4.2) to evaluate the required derivatives.*

Now the error function over all patterns E^n can be minimized with the help of an arbitrary minimization algorithm (e.g. gradient descent) using the derivatives of E with respect to w

$$\frac{\partial E}{\partial w_{ji}} = \sum_n \frac{\partial E^{(n)}}{\partial w_{ji}}. \tag{4.5}$$

As there is also an analogous technique to calculate the second-order derivatives efficiently with an algorithm similar to the backpropagation (Bishop, 1995, pp. 150–160), more complex minimization methods can be used (cf. section 4.3).

White (1989b) shows, that the estimator $\hat{\boldsymbol{w}}$ obtained by gradient descent backpropagation is consistent and asymptotically normal:

Theorem 4.22 (White, 1989b, p. 457)
Let $(\Omega, \boldsymbol{F}, P)$ be a complete probability space on which is defined the sequence of i.i.d. random variables $\{(X,Y)_t\} = ((X,Y)_t : \Omega \rightarrow \mathbb{R}^{d+k}, t = 1, 2, \ldots)$. Let $\ell : \mathbb{R}^{d+k} \times W \rightarrow \mathbb{R}$ be a function such that for each w in W, a compact

---

[76]The first step is often called the forward pass, while the combination of steps two and three is called backward pass.

subset of $\mathbb{R}^s$, $s \in \mathbb{N}$, $\ell(\cdot, \boldsymbol{w})$ is measurable-$\mathbb{B}^{d+k}$ and for each $(x, y) \in \mathbb{R}^{d+k}$, $\ell((x, y), \cdot)$ is continuous on W. Suppose that $\boldsymbol{w} \to \boldsymbol{w}^*$ a.s.-P where $\boldsymbol{w}^*$ is an isolated element of W^* interior to W. Suppose in addition that for each $(x, y) \in \mathbb{R}^{d+k}$ $\ell((x, y), \cdot)$ is continuously differentiable of order 2 on $\int W$; that $\mathbb{E}[\nabla((X, Y)_t, \boldsymbol{w}^*)^t \nabla \ell((X, Y)_t, \boldsymbol{w}^*)] < \infty$; that each element of ℓ is dominated on W by an integrable function; and that $A^* := \mathbb{E}(\nabla^2 \ell((X, Y)_t, \boldsymbol{w}^*))$ and $B^* := \mathbb{E}[\nabla \ell((X, Y)_t, \boldsymbol{w}^*) \nabla \ell((X, Y)_t, \boldsymbol{w}^*)^t]$ are nonsingular matrices. Then

$$\sqrt{n}(\hat{\boldsymbol{w}} - \boldsymbol{w}^*) \to^d \mathcal{N}(0, C^*) \; ,$$

where $C^* = A^{*-1} B^* A^{*-1}$. If in addition each element of $\nabla \ell \nabla \ell^t$ is dominated on W by an integrable function, then $\hat{C}_n^* \to C^*$ a.s.-P, where $\hat{C}_n^* = \hat{A}_n^{-1} \hat{B}_n \hat{A}_n^{-1}$, $\hat{A}_n^{-1} = \frac{1}{n} \sum_{t=1}^n \nabla^2 \ell((X, Y)_t, \hat{\boldsymbol{w}}_n)$ and $\hat{B}_n = \frac{1}{n} \sum_{t=1}^n \nabla \ell((X, Y)_t, \hat{\boldsymbol{w}}_n) \nabla \ell((X, Y)_t, \hat{\boldsymbol{w}}_n)^t$.

An equal result does also hold for misspecified models (Domowitz and White, 1982). Therefore gradient descent backpropagation is well-suited for calculating the LS estimator. Hypotheses tests of model fit can be calculated with the help of this asymptotic distribution of the estimates (cf. section 4.4).

4.2.2 Sequential Estimation with Backpropagation

Obviously, instead of using the sum in equation (4.5) the weights may be updated pattern-wise by

$$\boldsymbol{w}_{ji}^{(n+1)} = \boldsymbol{w}_{ji}^{(n)} - \eta_{n+1} \delta_j x_i \tag{4.6}$$

for the hidden and

$$\boldsymbol{w}_{kj}^{(n+1)} = \boldsymbol{w}_{kj}^{(n)} - \eta_{n+1} \delta_k z_j \tag{4.7}$$

for the output layer (Bishop, 1995, p. 146).

This is called online learning and we will call it sequential estimation of the parameters. Unfortunately, this is not a real online learning algorithm, as the solution obtained after one learning pass is not close enough to the (local) minimum and the algorithm has to iterate over the sample (Huang, Zhu and Siew, 2006). However, this sequential estimation procedure is asymptotically consistent.

White (1989c) pointed out that using the so called online backpropagation algorithm is a special version of the more general stochastic approximation method by Robbins and Monro (1951); Blum (1954); Chung (1954), which is applied to nonlinear LS regression.

Theorem 4.23 (White, 1989c, p. 1006)
For an i.i.d. sample $(X,Y)_n$ with $|(X,Y)_n| < \infty$ and $\{\eta_n \in \mathbb{R}^+\}$ a decreasing sequence such that

a) $\sum_{n=1}^{\infty} \eta_n = \infty$

b) $\lim_{n\to\infty} \sup(\frac{1}{\eta_n} - \frac{1}{\eta_{n-1}}) < \infty$ *and*

c) $\sum_{n=1}^{\infty} \eta_n^d < \infty$ *for some $d > 1$.*

Define the Backpropagation estimator

$$w^{(n)} = w^{(n-1)} + \eta_n(\nabla f(x^{(n)}|w^{(n-1)}))^t(Y^{(n)} - f(x^{(n)}|w^{(n-1)}))$$

with arbitrary $w^{(0)}$ and ∇f denoting the Jacobian. Then either
$w^{(n)} \to \{w : \mathbb{E}\left(\nabla(Y^{(n)} - f(x^{(n)}|w^{(n)}))\right) = 0\}$ *a.s. or* $w^{(n)} \to \infty$ *a.s. .*

Obviously, there is an additional algorithmic parameter to be chosen within the bounds mentioned in Theorem 4.23: η_n. This parameter can be calculated in a data-dependent[77] or independent way. The most simple choice would be $\eta_n = \frac{1}{n}$. This parameter has got two functions: It controls the numerical optimization algorithm, and it controls the influence of a single pattern on the target function. While the first property is given for batch and sequential estimation, the second property is only given for the sequential algorithm.

Dependence on Re-Orderings of the Sample Set and Time-dependent Modeling

Backpropagation has been developed and used for online learning in neural networks. Although the results of batch learning and online learning with backpropagation are approximatively identical, there are big differences in the finite case. First, it depends on the ordering of the sample. This is not wanted in modeling of a fixed relation between explanatory and response variables, but could be useful in modeling time-dependent relations.

---

[77]cf. section 4.3.

The updating mechanism depends on the relations between the variables of two "neighboring" observations and not on the time. Therefore, the "moving"-property is not time-dependent alone, but also data-dependent and no area of application for such a modeling is known to the author. Nevertheless, using the sequential algorithms provides a build-in adaptivity of the model to changes in the relationship; and this adaptivity is more influenced by changes in the data than by time.

Real Online Learning Capability

As mentioned earlier, the gradient descent algorithm suffers from slow convergence. Therefore, the data has to be processed various times even if sequential algorithms are used (Huang, Zhu and Siew, 2006). Thus, the sequential paradigm does *not* allow pure online estimation.

4.3 Numerical Contributions

Numerical optimization methods are algorithms that search the parameter space for optima. There are a lot of different optimization algorithms for nonlinear optimization problems in regression known today (Seber and Wild, 2003, pp. 587–660). Many of them have also been applied to MLP networks (Reed and Marks, 1998, p. 155–179).

In general, there are two major approaches: Local and global search. In local optimization, the algorithm starts from an arbitrary interior point of the parameter space. Depending on the surrounding of the actual point, it calculates a direction, which probably will improve the target function. Then, a new point in the calculated direction is chosen and a new direction is calculated. Local algorithms are performing well, if the starting point has been near the global optimum of the target function. They are classified by their use of derivatives of certain orders: There are algorithms using no derivatives at all, using only first-order derivatives and using second-order derivatives. In contrast, global optimization algorithms generate a sequence of points within the parameter space, check the values and keep the best solutions.

Obviously, these two approaches may be combined, for example by using more than one starting point for a local search. Furthermore, they can be combined by using the local information to generate new starting points at areas of the feature space, which have not been searched for an optimum so far.

The simplest first-order local search algorithm is *gradient descent*, where the direction of the local search is given by the negative gradient (Bishop, 1995). It has been improved by using momentum terms to overcome some problems with local search like oscillation and flat plateaus (Bishop, 1995; Duda, Hart and Stork, 2000), and using line search to estimate the best step size η. Second-order derivatives are used by conjugate gradients, Newton and quasi-Newton methods (Bishop, 1995). Concerning quadratic error functions the Levenberg-Marquardt algorithm (Marquardt, 1963) has been used successfully with neural networks (Hagan and Menhaj, 1994).[78]

Not every nonlinear optimization algorithm is well suited for incremental estimation. Obviously the simple gradient descent and conjugate gradients are directly implementable. Unfortunately, features like line search may cause problems because the algorithm probably will not converge in a stochastic setting. The stochastic local optimum is moving around the "real" optimum and only techniques, that use any decreasing parameter ensuring a reduction in approximation in each step, may converge (to any point "near" the deterministic local optimum). That is why only methods, which do not optimize the learning rate, are usable for stochastic settings in general.[79]

4.3.1 Numerical Complexity of Neural Networks

Using common terminology of the big $O(\cdot)$ notation[80], the complexity of neural networks can easily be calculated. Bishop (1995) shows that with W denoting the number of parameters (weights) in a neural network, the complexity of calculating and learning one pattern is $O(W)$.

Therefore, with N pattern the effort for online learning is $O(N \cdot W)$. In batch learning, the pattern set is G_1-times iterated[81], thus resulting in $O(N \cdot W \cdot G_1)$. The number of weights in a three-layered completely connected neural network

---

[78]Fine (1999) provides implementations of the gradient descent, the conjugate gradient, the quasi-Newton and the Levenberg-Marquardt algorithms for MATLAB®.

[79]Murata (1998), Saad and Rattray (1998) and Behera, Kumar and Patnaik (2006) suggest different methods for optimization of the learning rate η, that are useful within stochastic approximation.

[80]Graham, Knuth and Patashnik (1994) present a good overview of the big O notation.

[81]Obviously, $G_1 \in \mathbb{N}$ is stochastic due to the random initialization of the weights.

is dependent on the dimensions of input (D), of output (K) and the number of nodes in the hidden layer (M):

$$W = (K+1) \cdot M + (M+1) \cdot D = (K+D+1) \cdot M + D \in O((K+D) \cdot M)$$

Memory usage is also calculated and shown in Table 4.2; $G_2(K, D, M)$ denotes the space used by the numerical optimization algorithm.

Learning method	processing time	memory
online	$N \cdot (K+D) \cdot M$	$(K+D) \cdot M + G_2(K, D, M)$
offline	$N \cdot (K+D) \cdot M \cdot G_1(K, D, M)$	$(K+D) \cdot (M+N) + G_2(K, D, M)$

Tab. 4.2: Processor and memory usage

In general, if $G_1 < W \ll N$ can be supposed, the time complexity is linear in N for both learning schemes. Memory usage is constant for online learning in contrast to linear memory usage for batch learning.

4.3.2 Parallel Implementations of Multi-layer Perceptron Networks

Representing statistical models as neural networks provides a visual view on possible parallelization of the necessary calculations. This property is only discussed in brief here, the contributions by Sundararajan and Saratchandran (1998) and more recently Zhu and Sutton (2003) should be consulted to obtain more details on that topic.

Parallel Computation in the Learning Phase

In its forward pass, the backpropagation algorithm (cf. algorithm 4.1) calculates the output values as described in the production phase (see below). Besides this node parallelism, training set parallelism is a very flexible and efficient parallelism approach for batch learning. Its implementation is simple by adding the different weight updates by equation (4.5). Roughly speaking, the required time for batch learning a neural network scales with the factor $\frac{1}{\#P}$ where $\#P$ denotes the number of processors, which is bounded by the training set size.

Parallel Computation in the Production Phase

In contrary to the learning phase, parallel computation in the production phase is less important because of its lower computational effort. This is trivial node parallelization and demonstrated in Figure 4.4. However, the calculations required in the productive phase are also needed in the forward pass of the backpropagation in the learning phase. Therefore, even little time savings are valuable.

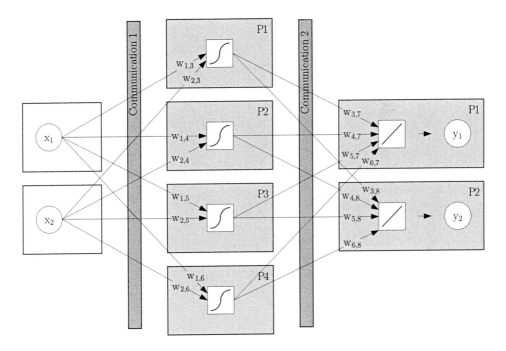

Fig. 4.4: Parallel scheme

Neglecting any communication effort, this approach to parallelization scales with factor $\frac{1}{\#P}$. It is obviously restricted by the maximum number of nodes per layer. This parallelization is very important, if we want to realize an online learning system because there are narrow time constraints to be met (cf. chapter 7).

4.4 Choosing the Appropriate Model Complexity

Choosing the right number of hidden nodes is difficult and there are a lot of different approaches in literature (cf. Ripley (1995) for some statistical approaches). Classical techniques like cross-validation are impracticable because we have to retrain different networks and should not use the same starting point or a subset of the fitted values as a new starting point (Ripley, 1996, p. 170f.).

In one approach, you start with a too large model and use regularization through adding penalizing terms which reduce the set of possible models (Fine, 1999, p. 215–221). This approach is not very appealing because there are too many free parameters, that reduce the degrees of freedom which results in a higher variance of the estimators. Early stopping algorithms also suffer from this shortcoming (Bishop, 1995, p. 343–345).

Other algorithms try to exclude unnecessary nodes from the net (Reed and Marks, 1998, pp. 219–235). This pruning is done by successively setting weights to zero and deleting a node which has only zero weights. Statistical tests are possible tools for decision to remove weights, and three popular test procedures for this task are described in subsection 4.4.1.

'Cascade correlation' is the most popular technique for growing a neural network during learning (Bishop, 1995, p. 357ff.). With each added node, the weights for the former added nodes are fixed and only the "new" weights are estimated. As this model is totally dependent on the ordering of the sample set, it should not be used.

4.4.1 Tests

In regression, statistical tests are used to select regressors or to choose an appropriate order for a polynomial regression function. They are used to prune too large models; but there are also techniques to find unattended nonlinearities (Seber and Wild, 2003, p. 228ff.).

LM Test

The Lagrange Multiplier (LM) test is one of the first tests proposed for the use with neural networks (White, 1989a). This test is used, to examine, if additional nodes would reduce the neglected nonlinearity significantly. Therefore, the hypothesis

$$\mathbb{H}_0 : P(\mathbb{E}(y|\boldsymbol{x}) = f(\boldsymbol{x}|\boldsymbol{\theta}_0)) = 1 \text{ for a } \boldsymbol{\theta}_0 \in \Theta \text{ versus}$$
$$\mathbb{H}_1 : P(\mathbb{E}(y|\boldsymbol{x}) = f(\boldsymbol{x}|\boldsymbol{\theta})) < 1 \text{ for all } \boldsymbol{\theta} \in \Theta$$

has to be tested.

If the model is correct, i.e. large enough to approximate the function f correctly, $\mathbb{H}_0$ would be true. Otherwise the probability, that the conditional expectation of the output y on the input $\boldsymbol{x}$ is given by the function f would be less than 1, i.e. there are differences between the function f and the expected values, outside the null set. This implies for neural networks, we would have to include $q \geq 1$ additional nodes in the hidden layer to the estimated network $f(\boldsymbol{x}|\hat{\boldsymbol{w}})$, i.e. the new function is

$$f(\boldsymbol{x}|\hat{\boldsymbol{w}}) + \sum_{a=h+1}^{h+q} \hat{w}_a^{(2)} g(\langle \hat{\boldsymbol{w}}_a^{(1)}, \boldsymbol{x} \rangle) \tag{4.8}$$

There are two possible concrete test procedures: We could test $\hat{w}_a^{(2)} = 0$, where $\hat{\boldsymbol{w}}_a^{(1)}$ is not defined under this null hypothesis, or we could test $\hat{\boldsymbol{w}}_a^{(1)} = \boldsymbol{0}$ and $\hat{w}_a^{(2)}$ would not be specified under the null hypothesis.
Concerning the first approach, White (1989a) suggests the use of carefully randomized values for the unidentified parameter. Unfortunately, this test procedure is not consistent, but yields good results in simulations (Lee, White and Granger, 1993).

The second approach based on the third order Taylor polynomial (Teräsvirta, Lin and Granger, 1993):

$$f(\boldsymbol{x}|\hat{\boldsymbol{w}}) + \langle \boldsymbol{\theta}, \boldsymbol{x} \rangle - \frac{1}{3} \sum_{i=0}^{I} \sum_{j=i}^{I} \sum_{k=j}^{I} \theta_{ijk} x_i \tag{4.9}$$

with $\theta_i = \sum_{a=h+1}^{h+q} \hat{w}_a^{(2)} \hat{w}_a^{(1)}$ and $\theta_{ijk} = \sum_{a=h+1}^{h+q} \hat{w}_a^{(2)} \delta_{ijk}$ with δ_{ijk} denoting the corresponding coefficients of the Taylor polynomial.

In both cases the test statistic is given by

$$LM = n\frac{\sum_n \hat{\hat{\epsilon}}_n^2}{\sum_n \hat{\epsilon}_n^2} \quad , \tag{4.10}$$

where $\hat{\epsilon}$ are the residuals from the function $f(\boldsymbol{x}|\boldsymbol{w})$ and $\hat{\hat{\epsilon}}$ are the residuals from the regression of $\hat{\epsilon}$ on (4.8) (with fixed $\hat{\boldsymbol{w}}_q^{(1)}$ or fixed $\hat{w}_q^{(2)}$ respectively). The statistic LM is asymptotically χ_q^2 distributed (White, 1989a).

Wald Test

Using the Wald statistic

$$n(R\hat{\boldsymbol{w}})^t(R\hat{C}R)^-1(R\hat{\boldsymbol{w}}) \sim \chi_q^2$$

where vector R selects the weights to be tested, q is the number of tested weights and $\hat{C}$ is the estimated covariance matrix.

Two popular variants of the Wald statistic have been proposed for the use with neural networks: Optimal Brain Damage (OBD) by Le Cun, Denker and Solla (1990) and Optimal Brain Surgeon (OBS) by Hassibi and Stork (1993); Hassibi, Stork and Wolff (1994). The variants differ in estimation of the correlation matrix C; the estimation is very rough in OBD as only the diagonal elements are estimated to simplify the calculation of the inverse. OBS uses the Hessian matrix instead (Ripley, 1997; Reed and Marks, 1998).

LR Test

Wald tests are large-sample equivalents of LR tests (Lehmann, 1983). Thus it seems to be advantageous to use the LR test directly, although this implies estimation of two models: The unreduced and the reduced model with estimated parameters $\hat{\boldsymbol{w}}$ and $\hat{\boldsymbol{w}}_r$ respectively (Anders, 1997). The LR test statistic is constructed by calculating the ratio of the likelihoods of the models ($L(\hat{\boldsymbol{w}})$ and $L_r(\hat{\boldsymbol{w}}_r)$)

$$LR = -2\ln\left(\frac{L(\hat{\boldsymbol{w}})}{L_r(\hat{\boldsymbol{w}}_r)}\right) \quad ,$$

which is χ_q^2 distributed.

Weights have to be removed singly in all three tests, as near-collinearity may lead to wrong decisions. Furthermore, the Wald test is known to have little power for large true values (Ripley, 1996, p. 169f.).

Unfortunately, these tests are based on residuals (LM and also the Durbin-Watson test (Refenes and Holt, 2001)), which are not available in real online application. They are based on crude estimations (OBD, OBS), or need two estimated models (LR test) to calculate the statistic.

4.4.2 Over-Fitting and Validation

As mentioned in section 2.4, MLP networks tend to over-fit the data, i.e. they additionally learn the errors, which are included in the learning sample. This results in estimated relationships which suffer from a too high complexity and poor generalization capabilities. The general approach of cross-validation is used with neural networks by Bishop (1995); Ripley (1996), thus its application is controversial (Fine, 1999). The major problems are the stochastic nature of the minimum search and the computational effort for calculating the cross-validation (Fine, 1999, p. 280f.).

Although other authors mention techniques like VC-dimension, cross-validation and bootstrapping, they suggest to use techniques of early stopping and weight decay to avoid over-fitting (Hastie, Tibshirani and Friedman, 2001). Both methods restrict the number of possible models and their complexity by preventing the models from being estimated completely. That means, the algorithm is stopped before it can reach the (local) minimum, or the parameter space is restricted by bounds which are more or less soft.

The method of early stopping is not very appealing as the model remains over-parameterized and the asymptotic of the estimators is bad, i.e. the estimated quadratic error of the parameter estimators is large. Weight decay really reduces the number of possible models (although the effect on the asymptotic of the estimators is not known completely) and empirical findings suggest that these techniques provide good results (Hastie, Tibshirani and Friedman, 2001, p. 356f.).

Weight decay is implemented by adding an additional term to the error function which is intended to describe the complexity of the model. As the complexity of a model may be described by the number of parameters, we could use $\langle w, w \rangle$ as penalty term which would be zero, if we used no parameters (Hastie, Tibshirani and Friedman, 2001, p. 356). Unfortunately this measure is dependent on the scaling of the input parameters. This is one of the reasons why the data should be standardized.

Using real test samples which are independent from the learning sample is a very useful alternative for large data sets. In the case of online learning, we could hold back the latest n_v observations from learning directly and save them in a cache for testing purposes. Every time a new observation arrives, the oldest observation from the test set is used for learning and the new observation is added to the test set. This scheme would ensure the independency of the learning and the test sample set in an online learning setting. Unfortunately, the most recent observations are not learned and the model is "out-of-date". This method seems to be appealing in settings, where adaption is less important and observations are "fast" collected relative to changes of the relationship.

4.5 Classification with Multi-layer Perceptron Networks

Although MLP networks are regression models, they have always been used for classification. MLP networks for regression and for classification differ in the choice of the transfer function for the output nodes: A linear transfer function is appropriate for regression and a sigmoidal type transfer function for classification. The logistic transfer function is usually proposed for classification. However, we can use any function with $\mathbb{R} \to [0, 1]$, because in nonlinear models the logistic link has lost its statistical interpretation (Cox, 1984).

At first glance, three different approaches to construct MLP networks for classification are plausible (cf. section 1.1). Taking a closer look reveals problems with the first approach, because coding k distinct values is complicated and not covered by standard techniques and implies a not existing order structure on the output data. Therefore, for general classification there are two approaches, whereas the latter may be further investigated and split up into two variants:

1) Classification
   Using k binary output nodes—one for each category—we have to ensure that exactly one is active at the same time, i.e. we need "cross-connections" between the output nodes.

2) Generalized Regression
   The binary decision lacks a lot of information about alternative decisions,

which might be useful, if the decision is narrow. There are two possible solutions with k output nodes taking values in the range $[0, 1]$:

a) Simple "regression"
   Using linear output nodes and the standard regression with target vectors in the 1-of-c coding scheme, i.e. a vector consisting of binary variables indicating the correct categories.

b) Affinities
   Using sigmoidal transfer functions, backpropagation (batch or sequential) with MSE can be used directly to "regress" on vectors of binary variables.

c) A-Posteriori Probabilities
   Using the softmax transfer function to construct a nonlinear variant of the GLM approach.

The first alternative is too complex for implementation because it would need threshold functions[82]. Therefore only the three variants of the second approach are described in more detail here.

Mean Squared Error and Simple Regression
For this very simple approach, the output is coded with the help an one-out-of-k classes coded target vector and standard regression is used (Bishop, 1995). It can be shown, that a network trained with this approach, approximates the optimal Bayes rule (Zhang, 2000, p. 453). As the target vectors satisfy the property of a probability function (the components of each target vector sum up to one), the sum of the estimated values is one (Bishop, 1995, p. 226). Nevertheless, the predicted values do not have to be positive, i.e. they formally cannot represent probabilities.

Calculation of Conditional Expectations
Learning a net with sigmoidal output nodes with the help of a differentiable error function (i.e. quadratic error) will construct a model, that calculates conditional expectations for the different classes (White, 1989b). Although it can be shown, that this learning scheme approximates the Bayes decision rule (Richard and Lippmann, 1991), the affinities do not always sum up to one.

Nevertheless, this combination of sigmoidal output and quadratic error

---

[82]This idea is realized by SVM in a more intuitive manner.

function can be estimated with the help of the standard backpropagation and therefore, it is very popular (Haykin, 1999).

Approximation of A-posteriori Probabilities

The most appealing approach is based on the same ideas like in GLM by using the softmax transfer function for the k output nodes (Bishop, 1995, p. 238):

$$y_r(z) = \frac{\exp(z_r)}{\sum_{r' \neq r} \exp(z_{r'})} \quad \forall r \in \{1, \dots, k\} \tag{4.11}$$

In this approach, the error function is usually defined as (Bishop, 1995, p. 237):

$$E = -\sum_n \sum_{r=1}^{k} t_r^{(n)} \ln\left(\frac{y_r^{(n)}}{t_r^{(n)}}\right) , \tag{4.12}$$

where $t^{(n)}$ denotes the target vector for pattern number n and $y^{(n)}$ denotes the vector of calculated category probabilities $P(y|x)$. Optimizing (4.12) yields the ML estimator (Bishop, 1995, pp. 236–240). Simulation studies and empirical analysis have shown, that this approach is superior to the MSE approach in performance and training time (Richard and Lippmann, 1991; Zhang, 2000).

This function has its minimum zero at $y_r^{(n)} = t_r^{(n)}$ for all n and is based on calculating the maximum likelihood estimator. Without standardizing its minimum to zero, we would have

$$E = -\sum_n \sum_{r=1}^{k} t_r^{(n)} \ln(y_r^{(n)}) . \tag{4.13}$$

This equation has the advantage, that $t_r^{(n)}$ is not included in the logarithm, but the minimum is greater than zero (cf. section 7.4).

Unfortunately, the model with k softmax output nodes is not identified. This can easily be seen because of the standardization to a probability distribution via the denominator of (4.11), there are different possible solutions for the vector z resulting to the same value for $t(\cdot)$.

We can archive unity, if we fix one of the predictor value $\eta_k = 1$ and use the response

$$y_r(\eta) = \frac{\exp(\eta_r)}{1 + \sum_{s=1}^{k-1} \exp(\eta_s)} \forall r \in \{1, \dots, k\} . \tag{4.14}$$

This can be seen as a nonlinear version of the GLM models. In Figure 4.5, the connection between GLM and MLP for classification is illustrated. The layer with linear transfer functions represents the predictors η used in GLM. It is obvious that this is a nonlinear extension of GLMs. The weights from the layer with the predictor values have the fixed value 1.

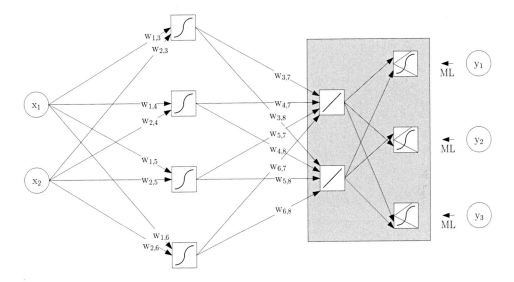

Fig. 4.5: Neural network for classification

This model is more restricted than the model with k softmax output nodes, as the number of parameters (for fixed number of nodes h at the hidden layer) is reduced by $h+1$, i.e. the number of incoming edges for the omitted redundant node. As the weights in the last layer have the fixed value 1, we can calculate the error function (4.12) in terms of the estimated predictors η_r:

$$
\begin{aligned}
E &= -\sum_n \sum_{r=1}^{k} t_r^{(n)} \ln \left(\frac{\exp(\eta_r^{(n)})}{t_r^{(n)} \left(1 + \sum_{s=1}^{k-1} \exp(\eta_s^{(n)})\right)} \right) \\
&= -\sum_n \sum_{r=1}^{k} t_r^{(n)} \left[\eta_r^{(n)} - \ln(t_r^{(n)}) - \ln \left(1 + \sum_{s=1}^{k-1} \exp(\eta_s^{(n)})\right) \right]
\end{aligned}
\tag{4.15}
$$

Although, the logistic response function—respective the softmax transfer function—is the connection between the natural parameter space of the multinomial distribution and the probabilities, it is not unique for nonlinear regression (Cox, 1984). Therefore, we could replace it by any other CDF (cf. section 7.4).

4.6 Ordinal Data in Multi-layer Perceptron Networks

MLP networks are nonlinear extensions of GLM and we can use the approaches from chapter 3 in neural networks. Therefore, we only describe some special properties that arise in modeling ordinal data in MLP networks. In particular, we need a constraint versions of the backpropagation algorithm.

4.6.1 Ordinal Input Variables

Subsection 3.1.2 deals with a method for ordinal input variables. This idea can be applied to MLP networks by adding an additional layer to map the ordinal values to a continuous variable. Figure 4.6 includes an example for the use of a simple coding scheme, a symmetric coding scheme, the additional layer and its weights.

With the help of the algorithm 3.1, we can solve (3.5), which follows a batch learning principle. This can easily be adopted for backpropagation neural networks with the help of an additional learning parameter describing the reduction factor for $\mu^{(k)}$ and repeating the batch learning until $\mu^{(k)} B(\boldsymbol{\alpha})$ is small enough.

Ordinal Data and Sequential Estimation

As MLP networks are estimated by sequential learning, we have to define a sequential learning algorithm for ordinal data.

Let us assume a three layer neural network with an additional layer for the ordinal coding of some input variables (cf. Figure 4.6). The estimators for the weights w_{kj} and w_{ji} are obtained via (4.6) and (4.7) as the derivative of the barrier function is zero for these weights. For all inputs i associated with ordinal values not the variable x_i is used but the calculated continuous variable z_i instead.

Now we can state:

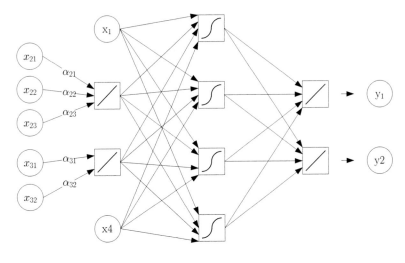

$$\alpha_{21} = \mathbb{1}\{x_2 = 2\}$$
$$\alpha_{22} = \mathbb{1}\{x_2 = 3\}$$
$$\alpha_{23} = \mathbb{1}\{x_2 = 4\}$$

$$\alpha_{31} = -\mathbb{1}\{x_3 = 1\} + \mathbb{1}\{x_3 = 4\}$$
$$\alpha_{32} = -\mathbb{1}\{x_2 = 2 + \mathbb{1}\{x_2 = 3\}$$

Fig. 4.6: Ordinal inputs of a MLP

Theorem 4.24

The weights α_{ij} are calculated by the following equations (for every i):

$$\Delta\alpha_{ic}^{(n+1)} = \begin{cases} \alpha_{ic}^{(n)} - \eta\left(\frac{c_i-2}{c_i-1}\zeta_c - \frac{1}{c_i-1}\sum_{\substack{\gamma=1 \\ \gamma\neq c}}^{c_i-1} \zeta_\gamma\right) & c \in \{1,\dots,c_i-2\} \\ 1 - \sum_{c=1}^{c_i-2}\alpha_{ic}^{(n+1)} & c = c_i - 1 \end{cases} \quad (4.16)$$

with

$$\zeta_c = \sum_{j=1}^{M} w_{ji}^{(n)} \delta_j x_{ic}^{(n)} - \frac{\mu^{(n)}}{\alpha_{ic}^{(n)}} \quad and$$

$$\delta_j = \frac{\partial}{\partial z_j}[g(z_j)] \cdot \sum_{k=1}^{c} w_{kj}^{(n)}\left(y_k - f(\boldsymbol{x}^{(n+1)}|\boldsymbol{w}^{(n)},\boldsymbol{\alpha}^{(n)})\right) \ .$$

Proof. Solving the Lagrange multiplier λ in the Lagrange system by minimizing f under $g = \sum_i \alpha_{ij} - 1 = 0$, we get

$$\lambda = \frac{\sum_i \frac{d}{dx_i} f(x)}{\sum_i \frac{d}{dx_i} g(x)} \ .$$

Here every derivation of g is constant 1 and thus we obtain

$$\frac{\sum_i \frac{d}{dx_i} f(x)}{c_i - 1} \ .$$

Together with (4.6) and (4.7), the result is given.   □

The parameter $\mu^{(n)}$ has to be reduced during learning (cf. Theorem 4.25 for details). This can be done by fixing a decreasing sequence of $(\mu^{(n)})_{n \in \mathbb{N}}$. Similar to the learning parameter η, this parameter has to be chosen carefully.

Additionally, we have to prove the consistency of this technique:

Theorem 4.25

Parameter estimates obtained by the constraint sequential backpropagation algorithm (algorithm 3.1 together with the updating rule 4.24) are either divergent, zero or consistent in the sense, that

$$w^{(n)} \to \left\{ w : \mathbb{E} \left[\nabla (Y^{(n)} - f(x^{(n)} | w^{(n)})) \right] = 0 \right\} \ .$$

Proof. We have to consider two cases.

case 1 $\mu^{(n)}$ is constant

Obviously, this is similar to standard sequential learning with backpropagation and the result is clear from Theorem 4.23.

case 2 The sequence $\{\mu^{(n)}\}_{n \in \mathbb{N}}$ decreases monotonically (cf. algorithm 3.1).

If parameter values diverge, the result is trivial because the influence of $\mu B(\alpha)$ decreases with growing components of α and so divergence is not influenced by μ. Let us assume that $(\theta^{*t}, \alpha^{*t})$ is the (local) minimum to which the algorithm will converge for a fixed μ from the actual parameter vector $(\theta^{*(n)}, \alpha^{*(n)})$. Two different cases for each component of the parameter vector α have to b distinguished:

a) $\nabla^2 \mu B(\boldsymbol{\alpha}^*) <^* \nabla^2 \frac{1}{2} \sum_n (y_n - f(\boldsymbol{x}_n|\boldsymbol{\theta}^*))^2$, i.e. $B(\boldsymbol{\alpha}^*)$ is dominated by $\frac{1}{2} \sum_n (y_n - f(\boldsymbol{x}_n|\boldsymbol{\theta}^*))^2$ in each direction

As $B(\boldsymbol{\alpha})$ is convex, reducing μ leads not to a very different minimum because the minimum is determined by the first part of the objective function (3.5) and only slightly translated by the second part, i.e. reducing μ does not change the local minimum significantly.

b) $\nabla_{\boldsymbol{\alpha}}^2 \mu B(\boldsymbol{\alpha}^*) > \nabla_{\alpha_i}^2 \frac{1}{2} \sum_n (y_n - f(\boldsymbol{x}_n|\boldsymbol{\theta}^*))^2$

In this case, the parameter tends to zero. The parameter μ with $B(\alpha)$ dominates the first part of objective function and the local minimum vanishes.

During the calculation, the second case could switch to the first case if μ is small enough. Nevertheless, one of these two possibilities is given asymptotically.

$\square$

4.6.2 Ordinal Output Variables

There are two approaches of ordinal MLP network classification known to the author.

First, the MLP variant of da Costa and Cardoso (2005) has got a single output node describing the probability parameter p of a binomial distribution of length k for k ordered classes. Obviously, this model suffers from a very restricted interpretation of ordinal data generation. The binomial distribution is usually interpreted as a series of k stochastically independent Bernoulli experiments with the same probability of success p. This interpretation suggests the use of this model for sequential data and to interpret p as the probability of reaching a higher class at a certain point of time. This model is very unrealistic because the probability of ascent is equal for every class (except for the last class, where it is zero). There are only a few situations in practice, where this property is given. Nevertheless, this model obviously ensures the unimodality of the conditional distribution $p(y|\boldsymbol{x})$. The used error function (da Costa and Cardoso, 2005)

$$\sum_{r=1}^{k} (f(r|\boldsymbol{x}) - \mathbb{1}\{r - y = 0\})^2 \ ,$$

where $y \in \{1, \ldots, k\}$ is the true class to which the object with features x belongs.

This is a quadratic error function and we can expect the model to approximate correctly the true conditional probability $P(r|x)$, if the model is correctly specified. The shape of the a-posteriori distribution is very fix and does not provide enough flexibility. Because of this and its restriction to a very narrow field of application this approach becomes unattractive.

Cheng (2007) introduces an alternative model, which is closely related to cumulative logit models (cf. section 3.3.1). This model uses k sigmoidal output nodes and is learned by k dimensional target vectors of the general form $t = (1, \ldots, 1, 0, \ldots, 0)^t$, where $t_i = 1$ for all $1 \leq i \leq y$ and $t_i = 0$ for all $y < i < k$. This is closely related to the approach of learning classifications by using MSE and sigmoidal outputs (cf. section 4.5). Therefore, it suffers from the same shortcomings. Additionally, the a-posteriori distribution is only unimodal by accident.[83]

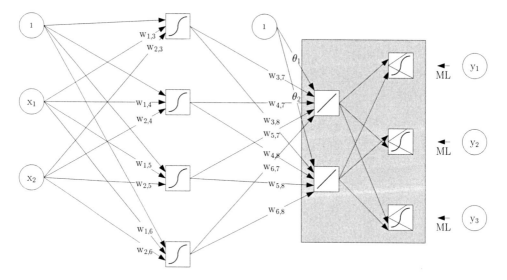

Fig. 4.7: Neural network for ordinal classification

Mathieson's model (1996), in which the linear predictors η_i of ordinal linear regression are replaced by nonlinear calculations, represents the most appealing approach. This model has to be completed by adding restrictions on parameters

---

[83]It is unimodal in most cases, but that is not ensured by the model.

in some special cases of ordinal output modeling (see Table 4.3). As these restrictions are usually made for the constant factor in the last layer, these "special" weights are denoted by θ_1 to θ_{k-1} (shown in Figure 4.7).

Referring to Table 3.3 and taking into account modeling techniques for neural networks, the Table 4.3 demonstrates the various possible approaches for modeling ordinal output values. As we have seen in section 4.6.1, we have to specify the relation

$$\boldsymbol{\pi} = h(\boldsymbol{\eta})$$

where $\boldsymbol{\pi}^t = (\pi_1, \cdots, \pi_k)$ and h denotes the response function. The used error function stays the same as in nominal classification (4.12). We only have to specify the appropriate transfer function and constraints.

model	transfer function(s)	constraints
cumulative	$\frac{\exp[\eta_r]}{1+\exp[\eta_r]} - \frac{\exp[\eta_{r-1}]}{1+\exp[\eta_{r-1}]}$	$\theta_1 < \cdots < \theta_{k-1}$
continuation-ratio	$\frac{\exp[\eta_r]}{1+\exp[\eta_r]} \prod_{s=1}^{r-1}\left(1 - \frac{\exp[\eta_s]}{1+\exp[\eta_s]}\right)$	$\theta_1 < \cdots < \theta_{k-1}$
adjacent-categories	$\frac{\exp(\sum_{s=r}^{k-1}\eta_s)}{1+\sum_{s=1}^{k-1}\exp(\sum_{s'=1}^{k-1}\eta_{s'})}$	$\eta_1 < \ldots < \eta_{k-1}$ (optional)
ordinal distribution	$\frac{\exp[\sum_{s=2}^{r}\eta_s]}{1+\sum_{s=1}^{k}\exp[\sum_{s'=1}^{s}\eta_{s'}]}$	$\eta_2 > \ldots > \eta_k$ (optional)

Tab. 4.3: Variants of link functions for ordinal regression

The models which are described in Table 4.3 and represented as a neural network structure in Figure 4.7 are generalized. They can be specialized to a nonlinear proportional odds model by reducing the number of nodes to one in the third layer.

Concave A-Priori Distribution

The optional constraints in Table 4.3 can be used to ensure unimodality of the a-posteriori distribution for every input vector $\boldsymbol{x}$.

Theorem 4.26 (Unimodality of a-posteriori distribution)
Assume the adjacent-categories or ordinal distribution model. Adding the constraints

$$\eta_1 < \ldots < \eta_{k-1} \text{ respective } 1 > \eta_2 > \ldots > \eta_k$$

leads to a concave a-posteriori distribution for all $\boldsymbol{x}$.

Proof. Obviously, for ordinal distribution holds:
The a-posteriori probabilities are ascending with r as long as $\eta_r > 0$. With $\eta_r < 0$, they are falling as $\sum_{s=1}^{r} \eta_s$ is falling. The argumentation is inverted for the adjacent-categories model.   □

This is directly given in proportional models with restrictions on the thresholds.

Corollary 4.27
Using a proportional-odds model, i.e.

$$\eta_r(\boldsymbol{x}) - \eta_s(\boldsymbol{x}) = c_{r,s}$$

is independent from $\boldsymbol{x}$ and additionally assuming the order on the thresholds

$$\theta_1 < \cdots < \theta_{k-1}$$

the a-posteriori distribution is unimodal.

Proof. The condition $\eta_1 > \eta_2 > \ldots > \eta_k$ is given. Theorem 4.26 provides the result.   □

5 Stochastic versus Batch Estimation

Supervised statistical learning can be performed by two major approaches: Sequential and batch estimation. While the properties of the batch learning principle—especially those of its estimators—have been researched in detail long ago, online learning still is a topic of research. Although there are a lot of results concerning the algorithmic and statistical properties of batch and sequential learning separately, these two aspects have never been researched concurrently.

Substantial work has been done for proving the asymptotically equivalence of batch and sequential learning principles (Saad, 1999). Nevertheless, the question arises in real applications, how efficient the approaches are in the finite sample case. Considering neural networks, efficiency cannot be measured in statistical or algorithmic terms separately.

A tradeoff between algorithmic (E_A) and statistical efficiency (E_S) has to be made. This tradeoff is often made in a qualitative way. Of course, these two aspects cannot be compared directly and the use of any weighting scheme like "$E = \lambda E_A + (1 - \lambda)E_S$" is not correct. The dimensions of these concepts are too different.

Many textbook authors on neural networks mention this problem without going into details. Stochastic estimation is often recommended on the grounds that observations in the sample are very repetitive (cf. Haykin (1999, p. 171f.), Duda, Hart and Stork (2000, p. 316) and Si, Nelson and Runger (2003, p. 47)). This argumentation is true, if the input variables are mainly categorical. Regarding parallelization, the batch estimation is superior to stochastic estimation because it can be used with pattern set parallelism (Haykin, 1999, p. 171f.) without any additional loss in accuracy.

Concentrating on statistic efficiency for fixed algorithmic effort is reasonable in context of neural networks. This requires calculation of finite sample

efficiencies of these learning concepts for practical purposes. It is also interesting
to fix the finite statistical efficiency and vary the algorithmic effort. Thus,
these two perspectives on finite sample statistical and algorithmic efficiency
have to be considered in analyzing algorithms.

However, batch and sequential estimation do not have to be different. For example, we can reformulate the batch/offline estimator to a batch-like sequential
estimator:

Example 5.1
The classical standard linear regression with design matrix D, parameter vector β and
homoscedastic, independent, unbiased errors ϵ

$$Y = D\beta + \epsilon$$

can be learned with the offline/batch principle (as it is a "degenerated" multi-layer
perceptron network) and the estimator is obtained by $(D^t D)^{-1} DY$. But forming the
corresponding sufficient statistics

$$\sum_n X_i^{(n)} Y^{(n)} \quad \forall i \quad \text{and} \quad \sum_n (Y^{(n)})^2$$

we can also easily construct a offline/batch variant of this estimator by storing these
statistics.

We can formulate the following statements for comparing batch-like online
and stochastic online learning:

1) The memory usage of batch-like online learning is higher or equal to stochastic online learning.

2) If the results are needed after each new pattern, the calculation effort in
   batch-like online learning is higher or equal to stochastic online learning.

3) The statistical efficiency in batch-like online learning is usually higher or
   equal to stochastic online learning, as it is invariant to permutations of the
   observations in the sample.

4) A higher degree of approximation can be obtained by using the batch-
   like online variant, if the solution has to be calculated by a numerical
   optimization algorithm.

5) Switching of the model (within an appropriate family) can be done easily in batch-like online learning without loosing the required information from passed data.

The third statement is quantified in the following sections of this chapter. The other statements will be discussed in chapter 7.

5.1 Sequential Estimation and Order Independency

Definition 5.2 (Langley, 1995)
A learner L is called sequential, *if L inputs one training experience at a time, does not reprocess any previous experiences, and retains only one knowledge structure in memory.*

Comment 5.3
MLP networks with online learning as well as the linear regression in example 5.1 are sequential learners. The statistics, that have to be stored in memory in example 5.1 additional to the parameter values belong to the same knowledge structure, as there is no set of data structures, which grows dynamically.

Definition 5.4 (Langley, 1995)
A learner L exhibits an order effect *on a training set of experiences T, if there exist two or more orders of T for which L produces different knowledge structures.*

Knowledge structures generated by a sequential learner for two different orders of training experiences as well as the influence of the order effect are shown in Figure 5.1.

This definition is restricted to a single training set, but we can transfer the property to the learner itself:

Definition 5.5 (Langley, 1995)
A learner L is order sensitive, *if there exists a training set T on which L exhibits an order effect.*

Comment 5.6 (Langley, 1995)
Similarly, a learner is order independent *if it never exhibits an order effect.*

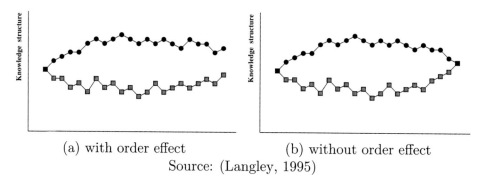

(a) with order effect   (b) without order effect

Source: (Langley, 1995)

Fig. 5.1: The knowledge structure generated by a sequential learner for two different orders of experiences

There are three levels of order effects because sequential learning can occur at three levels of temporal resolution: Concept, instance and attribute. While learning concepts is not the problem of data analysis, the other two levels are of high interest to data analysis. In this thesis, we deal with instance level sequential learning. Attribute based sequential learning might be interesting for data sets with huge patterns (see Figure 5.2).

The following example shows that order-dependence does not have to occur and that gradient descent leads to comparable estimation error in sequential and in batch learning:

A Simple Example: Estimation of the Mean

Let us suppose a sample of i.i.d. random variables $X_1, \ldots, X_n$. Although it is not usual, we can estimate the mean sequentially. Of course, it can be estimated sequentially with the help of the following well-known formula

$$\hat{\mu}_{i+1} = \frac{i}{i+1} \cdot \hat{\mu}_i + \frac{1}{i+1} X_{i+1} \; . \tag{5.1}$$

Suppose just for illustration that this formula would not exist. Further, there is no closed algebraic form of (5.1). In this case, we can use gradient descent with the updating rule

$$\tilde{\mu}_{i+1} = \tilde{\mu}_i - \eta_{i+1}(\tilde{\mu}_i - X_{i+1}) \tag{5.2}$$

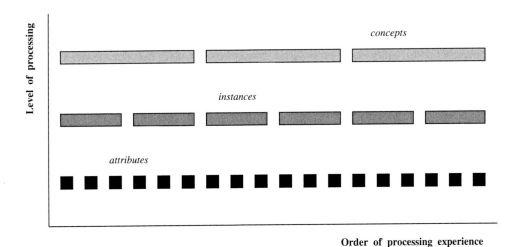

Source: (Langley, 1995)

Fig. 5.2: Three levels of temporal resolution in sequential learning

with starting value $\tilde{\mu}_0 = S$ and the (usually decreasing) learning rate $\eta_i \in \mathbb{R}$. This recursive formula can be reduced to a closed version:

$$\tilde{\mu}_n = \prod_{i=1}^{n}(1 - \eta_i) \cdot S + \sum_{i=1}^{n}\left(\eta_i X_i \prod_{j=i+1}^{n}(1 - \eta_j)\right) \tag{5.3}$$

It is known from stochastic approximation (Robbins and Monro, 1951) that η_i has to decrease in order to satisfy $\sum_{i=1}^{\infty}\eta_i^2 < \infty$ at least with order $\frac{1}{i}$ (Murata, 1998).

In order to obtain an order independent solution, we have to make sure that every random variable X_i has the same weight in this formula, i.e. $\eta_i \prod_{j=i+1}^{n}(1 - \eta_j)$ is constant. This is given, if $\eta_{i+1} = (1 - \eta_{i+1})\eta_i$. The sequence that satisfies this condition is unique with respect to its starting value: $\eta_{i+1} = \frac{1}{1+\eta_i}$.

Using $\eta_0 = \frac{1}{2}$ the closed form of the recursive formula (5.2) can be reduced to

$$\tilde{\mu}_{i+1} = \frac{1}{1 + i}S + \frac{1}{1 + i}\sum_{j=1}^{i}X_j \tag{5.4}$$

The expectation and the variance of the estimator $\tilde{\mu}_n$ for a sample of size n under the true mean μ are

$$\mathbb{E}_\mu(\tilde{\mu}_n) = \mu + \frac{1}{n}(S - \mu)$$
$$\mathrm{Var}_\mu(\tilde{\mu}_n) = \frac{1}{n} \cdot \mathrm{Var}_\mu(X)$$

$$(5.5)$$

This estimator is biased in relation to the difference between the starting value S and the true mean μ. This result is equal to that of the gradient descent algorithm for batch learning the mean. In the optimal case, i.e. if the mean is calculated directly with the known recursive formula (5.1), we obtain:

$$\mathbb{E}_\mu(\hat{\mu}_n) = \mu$$
$$Var_\mu(\hat{\mu}_n) = \frac{1}{n} \cdot Var_\mu(X)$$

$$(5.6)$$

Thus, the estimator is biased in the numerical gradient descent solution but with the same variance. As the bias tends to zero for large n, the estimators are asymptotically equivalent.

If we use batch learning with gradient descent based on

$$\tilde{\mu}_{i+1} = \tilde{\mu}_i - \eta_{i+1} \cdot \frac{1}{n} \sum_{j=1}^{n} (\tilde{\mu}_i - X_j)$$

we obtain

$$\mathbb{E}_\mu(\tilde{\tilde{\mu}}_n) = \mu + \frac{1}{n}(S - \mu)$$
$$\mathrm{Var}_\mu(\tilde{\tilde{\mu}}_n) = \left(\frac{n-1}{n}\right)^2 \frac{1}{n} \cdot \mathrm{Var}_\mu(X)$$

$$(5.7)$$

i.e. the variance is lower than in (5.6), but the bias is the same as in sequential online learning. This lower variance of the estimator "compensates" the bias and all in all the batch learning online variant is equivalent to the standard technique. Asymptotically, all three estimators are equal efficient.

This example shows that even in the case of simple estimation of an one dimensional parameter, there is a difference between the gradient descent in batch and in sequential learning. In both cases, it is biased in direction of the starting value; and the variance in sequential learning is equal to that of standard estimation, while it is lower in online batch learning. Furthermore, it demonstrates that order-dependence is not the only reason for a lower finite efficiency of the batch estimation procedure.

The calculation efforts of the direct estimator is $n + 1$, where we have to add n values and perform one division. The calculation effort of the sequential online estimation is $4n$, i.e. n steps with each consisting of two differences and two multiplications for calculation and usage of the learning rate. The batch learning is calculated with $n + 1$ operations for the calculation of the "mean" and $4n$ calculations for the stochastic approximation, i.e. $5n + 1$ in total. The complexities of all techniques belong to the linear class. This is possible, because we do not have to calculate the batch update for each iteration. If we neglect the linear structure, we would have to use $4n \cdot (n + 1) = 4n^2 + 4n$ operations, i.e. the batch estimation is of quadratic order for nonlinear optimization.

Memory usage (excluding temporary space) is 1 for the direct estimation and 2 for sequential estimation (actual value and learning rate).[84] Memory usage of the batch learning is either 2 if the linear structure is neglected or 3 otherwise ("statistic", actual value and learning parameter). The memory usage is constant for all methods.

These results cannot be transferred to more complex estimation problems, as the following example shows.

Example 5.7
Now we look at an example, which is two dimensional: Simple linear regression with one regressor. Besides batch learning with the well-known formulae

$$\hat{\beta}_1 = \frac{\sum_{i=1}^{n} x_i y_i - \frac{1}{n} \cdot \sum_{i=1}^{n} x_i \cdot \sum_{i=1}^{n} y_i}{\sum_{i=1}^{n} x_i^2 - \frac{1}{n} \left(\sum_{i=1}^{n} x_i\right)^2}$$

$$\hat{\beta}_0 = \frac{1}{n} \left(\sum_{i=1}^{n} y_i - \hat{\beta}_1 \sum_{i=1}^{n} x_i\right) ,$$

(5.8)

---

[84]The memory usage for storing the pattern data and the network is neglected for the moment. The usage of storage for the patterns is an important factor for comparison of online and offline learning though. In general, it is constant for online and linear for offline learning, of course.

we have to update the three values $\sum x_i$, $\sum y_i$ and $\sum x_i y_i$ for iterative learning. Stochastic approximation on the other hand is performed by:

$$\tilde{\beta}_{0,i+1} = \tilde{\beta}_{0,i} - \eta_{i+1} \left(\tilde{\beta}_{0,i} + \tilde{\beta}_{1,i} x_{i+1} - y_{i+1} \right)$$
$$\tilde{\beta}_{1,i+1} = \tilde{\beta}_{1,i} - \eta_{i+1} x_{i+1} \left(\tilde{\beta}_{0,i} + \tilde{\beta}_{1,i} x_{i+1} - y_{i+1} \right)$$

(5.9)

Obviously, it is not possible to obtain an estimator equivalent to the batch variant by stochastic approximation as long as there are three unknown quantities which are necessary for estimation.

These results can be generalized to every estimator, that can be estimated by stochastic approximation:
The batch variant and the sequential version of the estimators are asymptotically equivalent in the generalization error (Murata, 1998, p. 10); but the decreasing speed of errors is slightly slower for sequential estimation. Anyway, the sequential estimator has a much higher fluctuation resulting in a higher variance of the estimated variables (Murata, 1998, p. 8). In total, the performance of the batch estimation is better than that of the sequential estimation.

Heskes and Wiegerinck (1996) distinguish two subgroups of sequential learning: Cyclic and almost cyclic learning. In cyclic learning, the order of the observations is stable while iterating over the sample. In almost cyclic learning, the sample is stochastically reordered after each learning iteration. It is proven that almost cyclic learning performs better than cyclic learning. Similar results can be obtained for sequential batch learning which is located between sequential and batch learning (cf. section 5.2).

5.2 Gaining Order Independency for Sequential Estimators

There are also other techniques that avoid or reduce the ordering influence on the estimators provided by sequential estimation. The following approaches seem to be appropriate:

1. Choosing "good" orderings
   Constructing a good ordering of the sample seems possible. Nevertheless, it is a global optimization problem on the permutations of the sample,

i.e. it is inefficient. Local solutions only provide the global optimum by accident.

2. Averaging all estimators
   This is the most "statistical" idea: Estimate the parameters for every ordering of the sample and calculate their mean to obtain the order independent estimator. This procedure is theoretically supported by the Rao-Blackwell theorem (Lehmann, 1983). Unfortunately, the complete enumeration of all permutations of the sample is not practicable. Therefore, we could only use a subset of the permutations. In this case, the following additional questions have to be answered: How should we select a "good" subset of permutations? How many permutations should be used to obtain a sufficiently well-performing estimator?
   The Rao-Blackwell theorem suggests to use the permutations of the sample that provide estimators with the lowest possible correlation. In advance, we have to estimate these correlations to be able to choose the "right" permutations. Again, this is not practicable, as we would have to search all permutations. Considering MLP networks, there is another problem: If we stick into different local minima, averaging the estimators yields an unusable estimator. This is the reason why model averaging techniques like bagging are not usable for MLP networks.

3. Sufficient statistics
   Using sufficient statistics would solve various problems at once: First, they are order independent, i.e. every estimator that is only calculated with these statistics, would be independent from reordering the sample. Secondly, reducing the amount of data to be processed in higher order data analysis[85] provides various other applications, especially an user controlled relationship adaption.

The third idea is the basis for chapters 6 and 7. The other two ideas are discussed in the remainder of this section.

Optimizing the Ordering of the Observations
We have already seen, that it is impossible to search the whole space of possible orderings of the sample. Therefore, we have to optimize locally. This can

---

[85]i.e. data analysis based upon the statistics instead of the raw data.

be done by taking into account the observations learned so far to determine the "optimal" next observation. There are heuristics suggesting to choose that observation with the highest training error or that observation, which lies in a still uncovered part of the feature space (Haykin, 1999, p. 179). However, there are no published statistical results to support one of these heuristics.

Fukumizu (2000) has developed an algorithm, that provides statistical active learning by generating points in feature space, that should be learned next. Unfortunately, these points are generated independently from the sample and it is very likely, that generated points are not part of the sample. It has to be researched, if it is possible to use nearest neighbors to the calculated points within the sample instead of the calculated points themselves.

Sequential Batch Learning

Sequential batch learning is a hybrid approach between batch and sequential learning. There is a parameter n_b, which describes the size of a "cache" that is used to save the most recent observations until it is full. In this case, a single batch-like learning step is executed with the actual parameter estimation as the starting point and the observations in the cache as the learning sample.

The sequential batch learning principle includes the other two principles as special cases. It degenerates to sequential learning for $n_b = 1$ and it describes standard batch learning with $n_b = N$.

Unfortunately, there is no literature known to the author, that deals with the statistical analysis of sequential batch learning. This general analysis would allow to determine a good value for n_b, but this would only be of theoretical interest. n_b is chosen dependent on the space available at the processing unit for practical purposes.

5.3 Modeling of Time-Dependent Relationships

Time-dependent relationship modeling should not be mixed-up with time series analysis, although they are strongly related. Estimated relationships are not stable over time in reality. This means, any model, that has been constructed for decision support—like credit scoring models—have to be adapted to changes in the environment.

There are two major approaches to achieve this adaption: The standard approach and the online learning approach. The first approach can be regarded as the "batch" version of an adaptive system and the second as the corresponding sequential technique.

Standard Approach

The standard approach is characterized by collecting and storing huge amounts of data in databases for estimation of a new model based on the most recent data. After certain intervals of time, the model has to be revised, i.e. the estimators for the model are discarded and new estimators are calculated after re-weighting the items in the sample. The complete data has to be collected for this cause.

This approach suffers from a big drawback from the decision maker's point of view. Just before the old model is discarded and the new one is used, the decisions are based on old estimations. Furthermore, switching to the new model may result into inconsistencies in the decision process. For example, assume there is a credit scoring system, that (with the old model) would suggest to accept a credit application from a customer. The customer would get an offering, go home to think it over, come back the day after and would like to accept the offering. Now—with the new model—the system may not allow the credit anymore. In more mathematical terms, if the system changes continuously, the probability of situations alike can be minimized.

Online Learning

Online learning resolves this shortcoming. The parameter estimates are adapted during model use. However, using stochastic approximation with a fixed learning parameter can lead to high variations in the parameter estimates. This behavior is not desirable for practical purposes. We would like to control the "speed" of model adaption, i.e. to control the decrease in influence of the older observations[86]. This is possible with the help of sufficient statistics (cf. section 6.1.3).

---

[86]cf. Murata, Kawanabe, Ziehe, Müller and ichi Amari (2002); Murata, Müller, Ziehe and Amari (1997) for a first approach.

6 Sequential Statistics for Nonlinear Regression

One of the most important ideas of neural networks is the ability of analyzing large amounts of data by sequential learning. Sequential learning can also be realized with the help of sufficient statistics and thus will not suffer—in a statistical sense—from the shortcomings associated with Hebbian or backpropagation learning principles.

Additionally, parameter estimation in nonlinear regression is usually a nonlinear optimization task that can only be solved with an approximation algorithm. This often requires many iterations over the sample. Though, even if the data set can be stored completely, it has to be accessed and the equations have to be calculated in each iteration. Sufficient statistics reduce this effort in calculation. That is why statistics are very useful, if different models have to be estimated for the same sample, i.e. the amount of data to be processed can be reduced.

Nonlinear Regression

In general, nonlinear regression functions consist of functions that are not linear and for which no nonlinear transformation exists which would transform them to linear functions (Seber and Wild, 2003). In a more formal way, we define:

Definition 6.1 (nonlinear function)
A parametric function $f : \mathbb{R}^d \to \mathbb{R}$, $f(\boldsymbol{x}; \boldsymbol{\theta})$ *is* nonlinear, *if there is no bijective function* $g : \mathbb{R} \to \mathbb{R}$ *and there is no function* $\boldsymbol{p} : \mathbb{R}^d \to \mathbb{R}^d$ *with components* $p_i : \mathbb{R}^d \to \mathbb{R}$ *such that*

$$g \circ f(\boldsymbol{x}; \boldsymbol{\theta}) = \langle \boldsymbol{\theta}, \boldsymbol{p}(\boldsymbol{x}) \rangle \ .$$

It can be shown, that the LS estimator of the parameters of a nonlinear regression function has the same asymptotic properties as the LS estimator of the parameters for linear regression (Seber and Wild, 2003). Unfortunately, numerical minimization algorithms are needed to find the (local) minima. This is why we cannot be sure, whether the global optimum has been found. Therefore, the properties of the estimators are proven with respect to the set of local minima instead of the true parameter values.

6.1 Sequential Statistics for Linear Regression Models

Looking at the "normal" linear regression model

$$Y = X\beta + \epsilon \tag{6.1}$$

with LS estimation from a sample of size n, the parameters are found by

$$\beta = (X^t X)^{-1} X^t Y \ . \tag{6.2}$$

We can see from (6.2), that we have to invert the matrix $X^t X \in \mathbb{R}^{d \times d}$ and multiply two matrices in $\mathbb{R}^{n \times d}$. This can obviously be very time consuming for large n $(O(n^2) < O(n^?) \leq O(n^{2.376}))$ (Cormen, Leiserson and Rivest, 1990; Coppersmith and Winograd, 1987; Cohn, Kleinberg, Szegedy and Umans, 2005)[87]. All in all, calculating (6.2)—for large n—requires $O(n^2 \cdot d)$ operations.

Therefore, even such a simple algorithm like ordinary linear regression has to be calculated in a different manner for large samples. The following definition will help us to develop techniques with better algorithmic properties:

Definition 6.2 (Statistic for ordinary linear regression)
The statistic

$$T(X, y) = (\sum_{i=1}^n x_1^{(i)} x_1^{(i)}, \sum_{i=1}^n x_2^{(i)} x_1^{(i)}, \sum_{i=1}^n x_2^{(i)} x_2^{(i)}, \sum_{i=1}^n x_3^{(i)} x_1^{(i)}, \dots, \sum_{i=1}^n x_d^{(i)} x_d^{(i)},$$

$$\sum_{i=1}^n x_1^{(i)}, \dots, \sum_{i=1}^n x_d^{(i)}, \sum_{i=1}^n y^{(i)}, \sum_{i=1}^n y^{(i)} x_1^{(i)}, \dots, \sum_{i=1}^n y^{(i)} x_d^{(i)})$$

$$\tag{6.3}$$

is called the LS-sufficient statistic for linear regression.

---

[87]The value $O(n^{2.376})$ can actually only be obtained for quadratic matrices. More general, the famous Strassen algorithm which is usually used for large matrices has $O(n^{log_2 7})$.

The notation of (6.3) is unwieldy. Therefore, we use the following notation:

Definition 6.3
Let

$$T_{x_1^{p_1}\ldots x_d^{p_d} y^{p_d+1}} = T_{x_1^{p_1}\ldots x_d^{p_d}}(X,y) = \sum_{i=1}^{n} (x_1^{(i)})^{p_1} \cdots (x_d^{(i)})^{p_d} (y^{(i)})^{p_d+1}$$

denote a component of a statistic.

Now, we can refer to (6.3) with

$$T(X,y) = \left(T_{x_1^2}, T_{x_2 x_1}, T_{x_2^2}, T_{x_3 x_1}, \ldots, T_{x_d^2}, T_{x_1}, \ldots, T_{x_d}, T_y, T_{x_1 y}, \ldots, T_{x_d y} \right) \tag{6.4}$$

6.1.1 Properties of the Statistic

Definition 6.2 is not motivated out of statistical theory but out of the necessary calculations for the LS estimator. However, the term LS-sufficient can be very useful from an algorithmic point of view and it has to be defined more generally:

Definition 6.4 (LS-sufficient statistic)
Let $\boldsymbol{X} = (\boldsymbol{x}_1, \ldots, \boldsymbol{x}_n)$ be a i.i.d. sample from a statistical experiment $\mathcal{E} = (\Omega, \mathcal{A}, (P_\theta)_{\theta \in \Theta})$. Further let $\hat{\theta}'_{LS}(\boldsymbol{X})$ be the LS estimator for $\theta' \in \Theta' \subset \Theta$ from the sample $\boldsymbol{X}$ and $T : \Omega \to \mathfrak{X}$ be a measurable mapping for every $A \in \mathcal{A}$. $T(\cdot)$ is called LS-sufficient for Θ', if and only if there is a measurable function $f : \mathfrak{X} \to \Theta'$, such that

$$f \circ T(\boldsymbol{X}) = \frac{d}{d\theta'} LS_\theta(\boldsymbol{X})$$

for every $\boldsymbol{X}$ obtained from $\mathcal{E}$ and every $\theta' \in \Theta'$.

It is difficult to compare property 6.4 with the concept of sufficiency as we can see from the following example:

Example 6.5 (Uniform distribution)

It is well-known, that the ML estimator of the parameter a of the uniform distribution

$$f(x; a) = \begin{cases} \frac{1}{a} & 0 \leq x \leq a \\ 0 & x < 0 \text{ or } x > a \end{cases}$$

is $\hat{a}_{ML} = \max_i\{x_i\}$ and the sufficient statistic is

$$T(X) = \max_i\{x_i\} \ .$$

The parameter a can also be estimated (indirectly) via LS. The expectation value of the uniform distribution is $\mathbb{E}(X) = \frac{a}{2}$ and by solving

$$\min_a \overset{!}{=} \sum_i (\frac{a}{2} - x_i)^2$$

we obtain the LS estimator

$$\hat{a}_{LS} = \frac{2}{n} \sum_i x_i \ .$$

Thus, the statistic $\sum_i x_i$ is LS-sufficient and cannot be compared with the sufficient statistic.

The concept of LS-sufficiency can be used to analyze the real online learning capability of a statistic. Such a statistic is called online learnable, if it does not "grow" with growing n. As the estimator is based only on this statistic instead of the sample, it can be calculated online without any losses.

Now, we can state that the LS-sufficient statistic really *is* LS-sufficient:

Theorem 6.6

The statistic (6.3) is LS-sufficient.

Proof. Take the LS approach from (2.5) with linear f:

$$LS = \frac{d}{d\beta} \sum_i (\boldsymbol{\beta} x_i + \beta_0 - y_i)^2$$

$$= \frac{d}{d\beta} \left[\sum_{j,k} \beta_j \beta_k \sum_i x_{ji} x_{ki} + 2\beta_0 \sum_j \beta_j \sum_i x_{ji} + \beta_0^2 + 2 \sum_j \beta_j \sum_i x_{ji} y_i + \beta_0 \sum_i y_i \right]$$

$\square$

The calculation effort of this statistic is obviously restricted by $d^2 \cdot n$. We have to solve a linear equation system with $d + 1$ equations, i.e. we have to invert a $(d + 1) \times (d + 1)$ sized matrix. All in all, the algorithm using the statistic requires (for large n) $O(d^2 \cdot n)$ operations, i.e. we can exchange the exponents of d and n to obtain an algorithm whose calculation effort is linear in sample size.

Additionally, we can define minimality for LS-sufficient statistics:

Definition 6.7
A LS-sufficient statistic is minimal LS-sufficient, *if for every other sufficient statistic T' holds:*

$$T = h \circ T'$$

for an appropriate measurable function h.

Unfortunately, this characterization is very difficult to prove for a given statistic.

If the error term ϵ is completely specified, we can also prove the following result:

Comment 6.8
The LS-sufficient statistic 6.3 is sufficient for the regression model (6.1) in which the distribution of ϵ belongs to one of the following distributions: Normal, binomial, Poisson, negative binomial. It is not sufficient for gamma, beta, log-normal or double-exponential distributions of the error term. These distributions belong to the exponential family, i.e. they can be written in the form

$$f_X(\boldsymbol{x}; \boldsymbol{\theta}) = h(x) \exp \left(\sum_{i=1}^{s} \eta_i(\boldsymbol{\theta}) T_i(\boldsymbol{x}) - A(\boldsymbol{\theta}) \right) \quad , \tag{6.5}$$

where $T_i(\boldsymbol{x})$ denotes a sufficient statistic. The statistic T for the first four distributions is covered by (6.3). Furthermore, statistic 6.3 is known to be complete in the case of normal distributed errors (Lehmann, 1983).

6.1.2 Online Calculation of Some Important Characteristics and Test Statistics

Besides the online estimation, it should be possible to check the model and to calculate some popular test statistics online. Although online calculation of test statistics is usually not needed, it is required after the model has been fitted. Otherwise, it would be very expensive to access the whole sample again. Therefore, statistic-based calculation is preferable. It is demonstrated that many (but not all) tests and characteristics can be calculated with the help of this statistic.

Coefficient of Determination

The LS-sufficient statistic is also sufficient for calculating the coefficient of determination defined by

$$R^2 = \frac{\sum_{i=1}^{n} (\hat{y}^{(i)} - \bar{y})^2}{\sum_{i=1}^{n} (y^{(i)} - \bar{y})^2} \tag{6.6}$$

with $\bar{y} = \frac{1}{n} \sum_{i=1}^{n} y^{(i)}$ and $\hat{y}^{(i)} = x^{(i)} \beta$.

Therefore, using the statistic defined in (6.3) and with $\bar{y} = \frac{1}{n} T_y(X, y)$ equation (6.6) results in:

$$R^2 = \frac{\sum_{i=1}^{n} (\hat{y}^{(i)})^2 - 2\bar{y} \sum_{i=1}^{n} \hat{y}^{(i)} + n\bar{y}^2}{\sum_{i=1}^{n} (y^{(i)})^2 - 2\bar{y} \sum_{i=1}^{n} y^{(i)} + n\bar{y}^2}$$

$$= \frac{A(T(X,y)) - 2\bar{y} \left(n\beta_0 + \sum_{j=1}^{d} \beta_j T_{x_j}(X, y) \right) + n\bar{y}^2}{T_{y^2}(X, y) - n\bar{y}^2}$$

This shows that information in addition to statistic (6.3) has to be saved for calculation of the coefficient of determination.

The corrected coefficient of determination is calculated by

$$R_c^2 = R^2 - \frac{d \cdot (1 - R^2)}{n - d - 1} \tag{6.7}$$

and therefore it is only based on the coefficient of determination, input dimension and sample size.

Overall-F-Test

The Overall-F-Test is used for testing

$$\mathbb{H}_0 : \quad \boldsymbol{\beta} = \mathbf{0}$$
$$\mathbb{H}_1 : \quad \exists_j \beta_j \neq 0$$

based on the F-statistic

$$F = \frac{(n-d-1)R^2}{d(1-R^2)} \sim F_{d,n-d-1} \ . \tag{6.8}$$

As it is based on the coefficient of determination, its online calculation is trivial.

Significance of Parameters

Significance of parameters is usually tested with the help of the t-test of significance. For testing parameter j, i.e.

$$\mathbb{H}_0 : \quad \beta_j = 0$$
$$\mathbb{H}_1 : \quad \beta_j \neq 0$$

the statistic

$$t = \frac{\hat{\beta}_j}{\hat{\sigma}\sqrt{a_{jj}}} \sim t_{n-d-1} \tag{6.9}$$

is necessary, where a_{jj} denotes the j-th diagonal element of $(X^t X)^{-1}$. The value of a_{jj} can be obtained by using

$$X^t X = \begin{bmatrix} n & T_{x_1} & T_{x_2} & \cdots & T_{x_d} \\ T_{x_1} & T_{x_1^2} & T_{x_1 x_2} & \cdots & T_{x_1 x_d} \\ T_{x_2} & T_{x_1 x_2} & T_{x_2^2} & \cdots & T_{x_1 x_d} \\ \vdots & \vdots & \vdots & & \vdots \\ T_{x_d} & T_{x_1 x_d} & T_{x_2 x_d} & \cdots & T_{x_d^2} \end{bmatrix} \tag{6.10}$$

and the value of $\hat{\sigma}$ can be calculated with the help of statistic (6.3) as follows:

$$\hat{\sigma} = \frac{1}{n}\sum_i \left(\hat{y}^{(i)} - y^{(i)}\right)^2 - \left(\frac{1}{n}\sum_i \left(\hat{y}^{(i)} - y^{(i)}\right)\right)^2 \ .$$

The first part is directly related to the optimal value that has been calculated during LS estimation. The second summand consists of $\sum_i \hat{y}^{(i)} = \sum_i f(x_i|\boldsymbol{\beta}) = A(T(X,y))$ with appropriate linear function A and $\sum_i y^{(i)} = T_y(X,y)$. Thus, this test can also be calculated based on statistic (6.3).

Wald, Likelihood Ratio and Lagrange Multiplier Test

These tests require a ML estimation of the estimated parameters (Lehmann, 1986). Therefore, the semiparametric regression model (6.1) has to be completely specified, i.e. we have to add assumptions on the distribution of the error terms ϵ. Using a normal distribution with a parameterized covariance matrix is the most common choice:

$$\epsilon \sim \mathcal{N}(\mathbf{0}, \sigma^2 \mathbf{\Psi}(\boldsymbol{\theta}))$$

A regression model with the resulting parameter vector $\boldsymbol{\gamma} = (\boldsymbol{\beta}, \sigma^2, \boldsymbol{\theta})$ contains $c = d + h + 1$ parameters, where h denotes the number of parameters for the covariance matrix which is stable with respect to n (Judge, Griffiths, Carter Hill, Lütkepohl and Lee, 1985, p. 182).

The three tests are asymptotically equivalent, their advantages and shortcomings regarding their use in linear regression models are discussed by Judge, Griffiths, Carter Hill, Lütkepohl and Lee (1985, p. 184).

Obviously the LR test can be calculated simply with the help of statistic (6.3):

$$\lambda_{LR} = 2(L(\hat{\gamma}) - L(\hat{\gamma}_r)) \tag{6.11}$$

by using the estimators for the full $\hat{\gamma}$ and the restricted $\hat{\gamma}_r$ model. Therefore, we need a second estimation which additionally requires $O(c_r^{\log_2 7})$ operations for inverting the matrix of the restricted linear equation system plus $O(c_r^2)$ operations for the matrix-vector multiplication. The sufficient statistics do not have to be calculated again and therefore, the calculation complexity is independent of n.

Although the Wald and LM tests only require a single estimation, we have to calculate the information matrix $\boldsymbol{I}(\hat{\gamma})$ and some minor calculations in addition. Therefore, we can expect the calculation complexity of the LR test to be only slightly higher than that of the Wald and LM tests, but with a better power for finite samples (Judge, Griffiths, Carter Hill, Lütkepohl and Lee, 1985, p. 184).

White's Test on Homoscedasticity

Let ϵ denote the residuals from the regression of y on $\boldsymbol{x}$. The White test is based on the statistic (Judge, Griffiths, Carter Hill, Lütkepohl and Lee, 1985, p. 453)

$$TR^2 \sim \chi_A^2 \ , \tag{6.12}$$

where R^2 is the coefficient of determination for the regression of the residuals ϵ on the variables from $(\boldsymbol{x} \otimes \boldsymbol{x})$ (cross product) and an additional constant. A denotes the number of explanatory variables in this auxiliary regression. It can be calculated by $A = \frac{d(d+1)}{2} - 1$ in most cases.

Unfortunately, the LS-sufficient statistic (6.3) is *not* LS-sufficient for this auxiliary regression, because we are in need of some higher order moments of $\boldsymbol{x}$ resulting from the square used in the LS approach on the cross product.

Other Test Statistics

The described test statistics are some important examples for additional statistics that are necessary in the analysis of regression functions. Many statistics can also be calculated[88], but there are also statistics that can *not* be calculated based on statistic (6.3).[89]

6.1.3 Simulation of Stochastic Weight Approximation with Known Statistics

A decreasing learning rate is necessary to estimate stable relationships in neural networks. However, keeping the learning rate constant is very useful for tracking the relationship in evolving systems (White, 1989c). The learning rate is used for two purposes: It controls the approximation algorithm, and it also controls the decrease of influence of older observations. With the help of the sufficient statistic, this interrelationship can be relaxed.

It is plausible to use a decreasing weighting scheme for statistic (6.4) in estimating a changing relationship. For example, we can use an exponential type decreasing scheme with fixed parameter ρ:

$$T^{(n \leq t)}(\boldsymbol{X}, Y) = \rho T^{(n \leq t-1)}(\boldsymbol{X}, Y) + (1 - \rho) T^{(n=t)}(\boldsymbol{X}, Y) , \qquad (6.13)$$

where $T^{(n \leq t)}(\boldsymbol{X}, Y)$ describes the statistic (6.4) calculated for all observations up to the observation t and the summation is performed element-wise.

Example 6.9

A continuously changing simple regression function has been simulated by choosing the parameter $\beta = (2, 1)$ as starting point. The parameter has been changed by

---

[88]This is usually the case, if the statistic does not depend on the ordering of the sample.

[89]i.e. statistics based on recursive residuals (Kianifard and Swallow, 1996), e.g. test of nonlinearity (Seber and Wild, 2003).

$\delta\beta = (0.01, -0.01)$ after each 5th observation. 500 observations have been simulated with gaussian error $\mathcal{N}(0,1)$, i.e. the parameter vector for the last 5 observations has been $\beta = (2.99, 0.01)$.

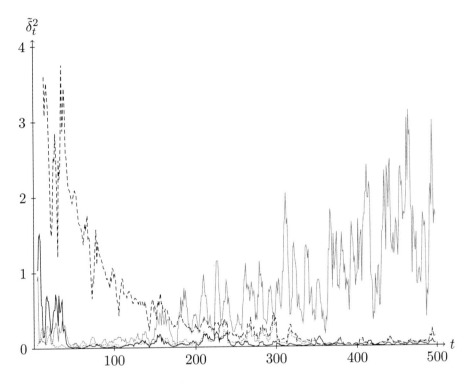

gray line: sequential standard regression,
black line: adaptive sequential regression,
dashed: stochastic approximation,
dotted: stochastic approximation with optimal starting values

Fig. 6.1: Moving regression function

Three different methods of estimation have been used: Sequential standard estimation (5.8), adaptive sequential estimation (5.8) with an adaptive statistic (6.13) using parameter $\rho = 0.96$ and stochastic approximation (5.9) with fixed learning parameter 0.01.

The following error function has been calculated to demonstrate performance of the estimation procedures:

$$\delta_t^2 = \left((\hat{\beta}_0 - \beta_0) + (\hat{\beta}_1 - \beta_1)x_t\right)^2 \tag{6.14}$$

In Figure 6.1, these values are plotted by smoothing them over 5 consecutive values with equal weights ($\tilde{\delta}_t^2 = \frac{1}{5} \sum_{t'=t-2}^{t+2} \delta_{t'}^2$).

Obviously, the standard estimation is not suited for estimating the parameters of a moving function. Depending on the starting value, the backpropagation may perform very well or very poor for the first observations. At the beginning, the adaptive statistic is in need of a certain amount of data to adjust the regression function, and it performs very well afterwards.

This approach does not need any model of a time series type. It only assumes a continuous change of the underlying function and therefore, it is nonparametric.

We have to choose the right parameter ρ to obtain a good approximation to the moving function at each time. Basically, this is the same problem like choosing a well-performing learning parameter for stochastic approximation.

6.2 Sufficient Statistics for Generalized Nonlinear Regression

The basic idea of sufficient statistics is reduction of information, which is provided by the sample, in order to simplify the analysis of the model parameters without loosing any necessary information. It is plausible that we have to separate the parameters from the single observation which is the fundamental idea behind the factorization criterion (Lehmann, 1983, p. 39).

For minimization purposes (used with ML and LS estimation techniques), we often need derivatives. The calculations are simplified a lot, if we use a linear function of the sufficient statistic and the model parameters. Taylor series expansions of nonlinear functions provide both properties: factorization of the target function in the case of LS estimation and easy calculation of the derivatives.

Let us have a look at the following special type of nonlinear regression functions:

$$f(\boldsymbol{x}|\boldsymbol{\theta}) = \theta_0 + \sum_{j=1}^{d} t_j[\theta_j x_j + \theta_{0j}] \;, \tag{6.15}$$

where t_j is a nonlinear function for every j with a Taylor series expansion for x on $\mathbb{R}$ with coefficients $\tau_{ji}(\boldsymbol{\theta})$.

This kind of nonlinear function is very special, but strongly related to neural networks, where the nonlinear parts are also functions from $\mathbb{R}$ to $\mathbb{R}$. The exponential function and the trigonometric functions are examples for the nonlinear parts t_j with on $\mathbb{R}$ convergent Taylor series expansions. The Probability Density Function (PDF) and the CDF of the normal distribution are also possible realizations of t because they are based on the exponential function.

The measurements of scale for explanatory and response variables in classical (non-)linear regression are metric. The following regression function with $f(\cdot)$ denoting the relationship between the variables is obtained (cf. subsection 2.1.3)

$$y = \ell_2^{-1} \circ f \circ \ell_1(x) + \epsilon \tag{6.16}$$

where $\ell_1(\cdot)$ and $\ell_1(\cdot)$ denote the scale systems.

If function (6.15) is used, both ℓ_2^{-1} and $\ell_1(x)$ collapse into the regression function.

6.2.1 Definition of a Set of Infinite Statistics

First, we have to define an infinite statistic:

Definition 6.10
Let $X \in \mathbb{R}^d$ be a random vector and $S = (X^{(1)}, \ldots, X^{(n)})$ a sequence of random vectors. Then an infinite statistic $T(X) \in \{\mathbb{R}^{n \times d} \to \mathbb{R}^\infty\}$ is defined by $T(X) = \{t_{i_1,\ldots,i_d}\}_{i_1,\ldots,i_d \in \mathbb{N}_0}$ with

$$t_{i_1,\ldots,i_d} = \sum_{n=1}^{N} \prod_{j=1}^{d} (X_j^{(n)})^{i_j} \ .$$

Comment 6.11

a) Definition 6.10 is meaningful because the function $S : (\mathbb{R}^{d \times n}, \mathfrak{B}^{d \times n}) \to (\mathbb{R}^\infty, \mathfrak{B}^\infty)$ is continuous and thus measurable.

b) The concepts of sufficiency and completeness are useful for infinite statistics without any changes in their formulation.

A set of statistics of this type is defined as follows:

Definition 6.12

Let $(X, Y) \in \mathbb{R}^d \times \mathbb{R}^k$ be a random vector and $S = ((X,Y)^{(1)}, \ldots, (X,Y)^{(n)})$ a sequence of random vectors. Let $T(X)$ be an infinite statistic and $\forall r \in \{1, \ldots, k\}$

$$T_r^\ell(X, Y) = \left(\sum_{n=1}^{N} \prod_{j=1}^{d} (X_j^{(n)})^{i_j} (Y_r^{(n)})^\ell \right)_{i_1, \ldots, i_d}$$

be another infinite statistic. Then

$$T = \{T(X), T(X,Y), T^2(X,Y), T^3(X,Y), \ldots\} \qquad (6.17)$$

is called a set of infinite statistics.

Comment 6.13

The asymmetry of X and Y reflects that the dimensionality of the infinite statistic is only dependent on the number of input variables d. In contrast, the number of elements in this set is also dependent on the used error function.

The definition (6.4) is a special case—a subset—of (6.17) as we can refer to (6.4) via $T = \{T(X), T(X,Y), T_s(Y), T_s^2(Y)\}$ for $s = 1, \ldots, d$.

A set of infinite statistics cannot be calculated. We have to reduce it to a finite one and we have to define a set of g-approximative infinite statistics.

Definition 6.14

Let $g \in \mathbb{N}$. Then the g-approximative infinite statistic $T(X|g) \in \{\mathbb{R}^{n \times d} \to \mathbb{R}^{g^d}\}$ to a infinite statistic $T(X)$ is defined by $T(X) = \{t_{i_1, \ldots, i_d}\}_{i_1, \ldots, i_d \in \mathbb{N}_0^{\leq g}}$ with

$$t_{i_1, \ldots, i_d} = \sum_{n=1}^{N} \prod_{j=1}^{d} (X_j^{(n)})^{i_j} .$$

Comment 6.15

Thus, $T(X|g)$ is the quadratic upper left sub-matrix $T(X)$ of size g. The parameter g controls the size of the set of statistics and the accuracy of the approximation of the Taylor series expansion (cf. subsection 6.2.3). In application, only a kind of triangle is required instead of the full square (cf. subsection 6.2.2).

The g-approximative set of statistics is not sufficient. But we can state, that the error, which results from the restriction of the statistic, converges to zero for $g \to \infty$.

Online Standardization of the Sufficient Statistic

The approximation error using a g-approximative statistic depends on the distance of the value that has to be calculated to the expansion point of the Taylor series, which is usually zero. This distance can be reduced by standardizing the variables. Besides this technical argument for standardization of the scales, sigmoidal-type transfer functions only "learn" well with small absolute input values (Daqi and Genxing, 2003). Using small parameter values as starting points and standardized response variables, we can maximize the probability that the parameters are estimated as fast as possible (Haykin, 1999, p. 181f.).

Generally speaking, the standardization is superfluous because it is a linear transform which can also be done by the parameters of input and output layers respectively.[90] Nevertheless, the parameters are also "standardized" because of standardized independent and dependent variables, i.e. they do not depend on the choice of scale. In this case, it is easier to choose good initial values for the numerical approximation algorithm.

Online standardization could be problematic because mean and variance of the different variables are not known a-priori, and afterwards, the different values of one variable are hidden within the statistic. Therefore, it is necessary to check, if the statistic can be "standardized" online. Let $\check{x}$ denote the standardized x that is

$$\check{x} = \frac{nx - \sum_i x^{(i)}}{\sqrt{n \sum_i \left(x^{(i)}\right)^2 - \left(\sum_i x^{(i)}\right)^2}}$$

and $\check{T}$ is the statistic of the standardized values.

---

[90]Otherwise, the model would be wrong, as it is not independent from linear transformations of the scale which is required for metric variables (cf. subsection 2.1.3).

Now, we have to calculate $\check{T}$ with the help of T.

$$\check{T}_{x_j^k} = \sum_{i=1}^{n} \left(\check{x}_j^{(i)}\right)^k = \frac{\sum_{i=1}^{n}\left(nx_j^{(i)} - T_{x_j}\right)^k}{\left(\sqrt{nT_{x_j^2} - \left(T_{x_j}\right)^2}\right)^k} = \frac{\sum_{\ell=0}^{k}\left(\binom{k}{\ell}n^\ell\left(-T_{x_j}\right)^{k-\ell}T_{x_j^\ell}\right)}{\left(\sqrt{nT_{x_j^2} - \left(T_{x_j}\right)^2}\right)^k}$$

For simplifying the equations, let

$$\zeta_{\ell,k_j}(T) = \binom{k_j}{\ell}n^\ell(-T_{x_j})^{k_j-\ell},$$

then, with the help of some algebra, the standardized statistic

$$\check{T}_{\prod_{j=1}^{d} x_j^{k_j}} = \frac{\sum_{\forall_j \ell_j \in \{0,\ldots,k_j\}}\left(\prod_{j=1}^{d}\zeta_{\ell_j,k_j}\right) \cdot T_{\prod_{j=1}^{d}\left(x_j^{(i)}\right)^{\ell_j}}}{\prod_{j=1}^{d}\left(\sqrt{nT_{x_j^2} - \left(T_{x_j}\right)^2}\right)^{k_j}}$$

can be calculated from T. An analogues extension of this equation can be used to calculate the values for $T_{y^{k_0}\prod_j x_j^{k_j}}$ and is omitted here.

Online standardization is possible and should be used to reduce approximation errors and to speed up convergence of the optimization algorithm.

6.2.2 Statistical Properties of the Set of Infinite Statistics

Statistically speaking, a good statistic should have two properties: First, it has to aggregate a data set to reduce the amount of information that is *sufficient* for a special task—usually the estimation of model parameters. Secondly, this data reduction should be done as far as possible, i.e. the statistic should be *minimally sufficient*.

The statistic defined in Definition 6.12 is minimally sufficient, as it is sufficient and complete in the exponential family. We will concentrate on the two most commonly used distributions: The normal distribution for standard regression and the multinomial distribution for categorical regression.

Standard Regression

Let $Y \in \mathbb{R}$ be a random vector and $(\boldsymbol{x}, \boldsymbol{Y}) = ((x, Y)^{(1)}, \ldots, (x, Y)^{(n)})$ an i.i.d. sample for a regression model with regression function $Y = f(x|\theta) + \boldsymbol{\epsilon}$ with the (fixed) points $(x^{(1)}, \ldots, x^{(n)})$, $x^{(i)} \in \mathbb{R}^d$ and the parameter vector $\theta \in \mathbb{R}^p$. Let the errors $\boldsymbol{\epsilon} = (\epsilon^{(1)}, \ldots, \epsilon^{(n)})$ be unbiased, homoscedastic, uncorrelated and be normal (for the moment), i.e. like in classical regression.

Using a function f like the one in (6.15), we can state:

Theorem 6.16

Let there be a convergent Taylor series of f in the range of x_i for all i with coefficients $\tau_t(\theta)$ and $\tau_t \neq 0$ for all $t \in \mathbb{N}$. Then the distribution of $(Y|x)$ belongs to the exponential family with statistic (6.17) without $T(x)$.

Proof. The probability distribution of $(Y|x)$ for normal errors is calculated as follows:

$$P^{Y|\boldsymbol{x}}(y|\boldsymbol{\theta}) = \frac{1}{\sqrt{2\pi\sigma^2}} \exp\left[-\frac{(y - f(\boldsymbol{x}|\boldsymbol{\theta}))^2}{2\sigma^2}\right]$$

$$= \frac{1}{\sqrt{2\pi\sigma^2}} \exp\left[-\frac{y^2 - 2\sum_{i=1}^d \sum_{t=0}^\infty \tau_{it}(\boldsymbol{\theta})yx_i^t + \left(\sum_{i=1}^d \sum_{t=0}^\infty \tau_{it}(\boldsymbol{\theta})x_i^t\right)^2}{2\sigma^2}\right]$$

With the Cauchy product for series we get

$$\left(\sum_{i=1}^d \sum_{t=0}^\infty \tau_{it}(\boldsymbol{\theta})x_i^t\right)^2 = \sum_{i,j}^d \left(\sum_{t=0}^\infty \tau_{it}(\boldsymbol{\theta})x_i^t\right)\left(\sum_{t=0}^\infty \tau_{jt}(\boldsymbol{\theta})x_j^t\right)$$

$$= \sum_{i,j}^d \sum_{t=0}^\infty \sum_{\ell=0}^t \tau_{i,\ell}(\boldsymbol{\theta})\tau_{j,t-\ell}(\boldsymbol{\theta})x_i^\ell x_j^{t-\ell}$$

Now, looking at an i.i.d. sample we have the probability density

$$P^{(Y|\boldsymbol{x})_n}(\boldsymbol{y}|\boldsymbol{\theta}) = \prod_{i=1}^{n} P^{(\boldsymbol{x},Y)^{(i)}}(y^{(i)}|\boldsymbol{\theta})$$

$$= \frac{1}{\sqrt{2\pi\sigma^2}} \exp\left[-\frac{(y^{(i)})^2 - 2\sum_d \sum_{t=0}^{\infty} \tau_{dt}(\boldsymbol{\theta}) \sum_i^n y^{(i)} (x_d^{(i)})^t}{2\sigma^2}\right]$$

$$\cdot \exp\left[-\frac{\sum_{d,d'} \sum_{t=0}^{\infty} \sum_{\ell=0}^{t} \tau_{d,\ell}(\boldsymbol{\theta}) \tau_{d',t-\ell}(\boldsymbol{\theta}) \sum_i^n (x_d^{(i)})^\ell (x_{d'}^{(i)})^{t-\ell}}{2\sigma^2}\right]$$

This can be mapped

$$P^{(Y|\boldsymbol{x})_n}(\boldsymbol{y}|\boldsymbol{\theta}) = \exp\left[-\eta(\boldsymbol{\theta})^t T(Y) - A(\eta(\boldsymbol{\theta}))\right] \quad ,$$

which is the general form of the exponential family.   □

Comment 6.17

Because the value of $T(X)$ is theoretically known a-priori, it is no random variable and thus not included in the statistic in the narrow sense. In practice—especially in social science—the values of the regressors are not known a-priori. Fortunately, the estimation procedure does not change, if stochastic regressors are allowed. Thus, we have to add $T(X)$ to the set of infinite statistics to get the necessary values.

Comment 6.18

Higher order interactions between the explanatory variables are not needed (cf. Theorem 6.16). Therefore, we only need the upper left triangle of the g-approximate statistic, what reduces the number of values to be stored for the sufficient statistics to

$$O\left(\binom{2g+d-1}{2g}\right) \quad . \tag{6.18}$$

This results from the interpretation of taking $2g$ variables out of d possible variables with repetition.

Theorem 6.19

Statistic (6.17) is sufficient and complete for estimating the parameter vector θ in the described nonlinear regression model with normal error.

Proof. Theorem 6.16 and sufficient statistics that are of full rank in exponential families are minimal sufficient (Lehmann, 1983). □

Because minimal sufficiency follows from sufficiency and completeness of a statistic, we can state that further data aggregation cannot be done without any losses in information necessary for estimation.

Comment 6.20
Theorem 6.19 holds for the infinite statistic, if and only if $\tau_t \neq 0$ for all $t \in \mathbb{N}$. If a finite number of τ_t is equal to zero, the infinite statistic may be reduced by some single entries, but its size stays infinite. But if there exists a value T with $\tau_t = 0$ for all $t > T$, then the statistic is also bounded. In this case, the function degenerates to a polynomial.

Although these results do not help us to obtain a better estimator for the parameters in nonlinear regression, the Lehmann-Scheffé theorem states that every unbiased estimator (like the LS estimator) based on this statistic is uniformly best (Lehmann, 1983). The influence of the summands of the Taylor series expansion on the value of $f(x|\theta)$ converges towards zero as $t \to \infty$. We can suppose that the higher order terms of the series can be omitted, as long as we do not leave a certain area around the expansion point. The loss of information in limiting the series will depend on the concrete form of the series, i.e. on the form of the used regression function.

As we usually do not provide any assumptions on the distribution of the errors (unless unbiasedness and homoscedasticity), we have to use the LS estimation technique instead of ML, and the concept of sufficiency is not applicable.

Comment 6.21
The statistic from Definition 6.10 is LS-sufficient for every regression function f of the form (6.15). The proof is straightforward.

Categorical Regression

In contrast to the situation of Theorem 6.16, we now have a regression on the parameters of a binomial or multinomial distribution respectively. Using the logit link function (4.14) (cf. subsection 2.3.3) and f to construct a generalized nonlinear regression model, we can state:

Theorem 6.22 (multinomial logistic regression)
Let there be a convergent Taylor series expansion of f_r in the range of x with coefficients $\tau_t(\theta)$ and $\tau_t \neq 0$ for all $t \in \mathbb{N}$, $f = (f_1, \ldots, f_{k-1})$. Then the distribution of $(Y|x)$ belongs to the exponential family with the (set of infinite) statistics

$$\left(T_1^1(X,Y), \ldots, T_{k-1}^1(X,Y) \right) \ .$$

Proof. The proof is analogue to the proof of Theorem 6.16. The probability for a sample $((x,y)^{(n)}, \ldots, (x,y)^{(n)})_n$ is

$$P(y|X,\theta) = \prod_n \exp \left[\sum_r^{k-1} f_r(x^{(n)}) t_r^{(n)} + \log \left(1 + \sum_r^{k-1} \exp \left(f_r(x^{(n)}) \right) \right) \right]$$

As the part with the logarithm does *not* contain any random variables, it is sufficient to have a look at the sum $\sum_r^{k-1} f_r(x^{(n)}) t_r^{(n)}$ which can be transformed like in proof to Theorem 6.16.   □

In contrast to regression, this result cannot be transferred to stochastic x as it involves the logarithm which cannot be represented by a power series expansion on the positive real line. Therefore, we are in need of other approaches that are not based on the binomial or multinomial distribution.

This is possible for probit classification.

Theorem 6.23 (general categorical regression)
Let there be an absolutely convergent Taylor series expansion of f_r in the range of x with coefficients $\tau_t(\theta)$ and $\tau_t \neq 0$ for all $t \in \mathbb{N}$. There is a power series expansion for the response ℓ on $\mathbb{R}$ (e.g. given for the probit). Then the distribution of $(Y|x)$ belongs to the exponential family with statistic (6.17).

Proof. The probability for a sample $((x, y)^{(n)}, \ldots, (x, y)^{(n)})_n$ using the link function ℓ is

$$P(y|X,\theta) = \prod_n \prod_r^{k-1} \ell(f(x^{(n)}|\theta))^{t_r^{(n)}} \cdot \left[1 - \sum_r^{k-1} \ell_r(f(x^{(n)}|\theta))\right]^{t_r^{(n)}}$$

Forming the log-likelihood

$$\log P(y|X,\theta) = \sum_n \sum_r^{k-1} t_r^{(n)} \log\left[\ell(f(x^{(n)}|\theta))\right] + t_k^{(n)} \log\left[1 - \sum_r^{k-1} \ell(f(x^{(n)}|\theta))\right]$$

There are power-series representations for $\ell(\cdot)$ on $\mathbb{R}$ and for $\log(\cdot)$ on $(0, 1)$. A power series representation is obtained by some algebra.   □

Comment 6.24

a) *This result also holds for all response functions with a power series representation on* $\mathbb{R}$.

b) *The logit response does not have a power series expansion on* $\mathbb{R}$, *especially* $1 + \exp(x)^\alpha$ *is only convergent on* $(-e, e)$.
   Nevertheless, we can use the probit response because the link function does not provide any special interpretation in nonlinear models (Thompson and Baker, 1981; Cox, 1984).

   More precise calculations are provided for the special case of MLP networks in chapter 7.

6.2.3 Approximation Accuracy

The choice of the parameter g which describes the finite approximation of the Taylor series expansion is crucial because it determines the amount of memory required for storing the statistic. Taking a too low value for g will result into a model, in which estimators either diverge or converge to small values resulting in a too smooth regression function.

The following example illustrates this problem:

Example 6.25

Let us assume a relationship between two interval-scaled variables X and Y is represented by

$$f(x) = \beta_0 + \beta_1 \exp[\beta_2 x] \qquad (6.19)$$

with appropriate parameter $\boldsymbol{\beta} = (\beta_0, \beta_1, \beta_2)^t$. We have already seen in subsection 2.1.3 that this function represents a feasible relationship between two interval-scaled variables. We have to solve for LS estimation

$$\left(\sum_{i=1}^{n} f(x_i | \boldsymbol{\beta}) - y_i \right)^2 \overset{!}{=} \min_{\boldsymbol{\beta}}$$

i.e. we have (in series representation)

$$E_n = \sum_{i=1}^{n} (f(x_i | \boldsymbol{\beta}) - y_i)^2 = n\beta_0^2 + 2\beta_0\beta_1 \left[\sum_{k=1}^{\infty} \frac{\beta_2^k}{k!} \sum_{i=1}^{n} x_i^k \right] + \beta_1^2 \left[\sum_{k=1}^{\infty} \frac{(2\beta_2)^k}{k!} \sum_{i=1}^{n} x_i^k \right]$$

$$- 2 \left[n\beta_0 \sum_{i=1}^{n} y_i + \beta_1 \sum_{k=1}^{\infty} \frac{\beta_2^k}{k!} \sum_{i=1}^{n} x_i^k y_i \right] + \sum_{i=1}^{n} y_i^2$$

The data is metric in this example. Thus, we can standardize both, explanatory and response variables. This leads to a smaller influence of uneven moments and smaller values in both parameters and variables during calculation. Therefore, it also reduces the error made by using the Taylor series approximation of the exponential function. We use the following weight decay on β_1 to compensate the numerical problems, that occur by leaving the area of good approximation:

$$2n \frac{(2\beta_1)^{g*}}{(g*)!} \text{ with } g* = 2\lceil \frac{g+1}{2} \rceil$$

This equation is motivated by the fact that

$$2 \frac{|x|^{g+1}}{(g+1)!} \text{ for } |x| \le 1 + \frac{1}{2}g$$

holds for the error made by approximating the exponential function with a polynomial of degree g. The penalty approximates zero for $g \to \infty$.

We have simulated 200 data points from function (6.19) in the range $[-1, 1]$ with additive Gaussian error with zero mean and standard deviation of one. These data have been used to fit the regression function (6.19) with and without the help of the approximate sufficient statistics (see Figure 6.2).

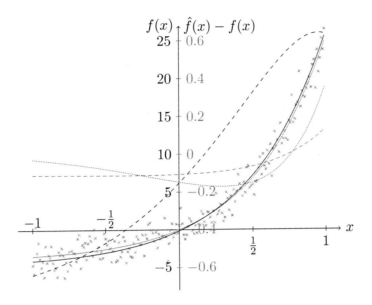

gray line: function, crosses: simulated data,
black line: estimated function based on 6th-approx. statistic
dashed gray: error of direct estimation (without statistic),
dashed black: error of estimation based on 6th approx. statistic,
dotted: error of estimation based on 10th approx. statistic

Fig. 6.2: Graphs of the function and its approximations

This example shows explicitly that estimation of parameters in a nonlinear regression context is possible with the help of the introduced sufficient statistics and that setting $g = 6$ is sufficient (in this example).

6.2.4 Determination of g and the Bias-Variance-Time-Memory Tradeoff

The right value for g is difficult to find. You will waste computational power and memory, if the value for g is too large. In contrast, if you choose a too small value, the optimization algorithm will probably diverge. Using weight decay as a penalty term can be used to ensure convergence of the algorithm, but we can expect that the estimators are biased to some extend. Therefore, we will have to enlarge the model complexity to be sure that the estimators are less biased. This results in a higher variance of the estimators.

This tradeoff is a bias-variance-time-memory one which is closely related to the statistical bias-variance tradeoff and the algorithmic time-memory tradeoff.

Determination of the right size of g is highly correlated to the used type of (regression) function and the range of the input and output variables. Therefore, it has to be calculated for every special application separately.

The Taylor series expansion has got a very problematic property: It is a (multi-dimensional) polynomial in the input variables. This means that it approximates infinity quickly at both ends of a certain interval around zero. Therefore, we have to ensure that the values do not lie outside of this interval.

This can be achieved by using a weight decay penalty term like it has been used in example 6.25. The used penalty

$$n\frac{|2\beta_1|^{g+1}}{(g+1)!} \tag{6.20}$$

is specific for the exponential because it is based on the remainder term in its Lagrange form (see Figure 6.3).

In its Lagrange form, the remainder term of a Taylor series expansion around zero is given by

$$R_g = \frac{f^{(g+1)}(\zeta)}{(g+1)!}x^{g+1} \tag{6.21}$$

for a special $0 < \zeta < x$. This can be used to prove the following theorem:

Theorem 6.26
Let $E(\boldsymbol{\theta}|x)$ be the quadratic error function for regression on a function $f : \mathbb{R} \to \mathbb{R}$ with Taylor series approximation $T(x)$ up to degree g. Let $R_g(x)$ denote the Lagrangian remainder. If $\max_{k\leq\lambda}\{f^{(k)}(0)\} < \infty$, it is true:

$$E(T^{(g)}(X,y)) + \max_{k\leq\lambda}\{f^{(k)}(0)\} \cdot n \cdot R_g(6 \cdot \max_n|\tilde{x}_n|) \geq E(\boldsymbol{\theta}|x)$$

for $g \geq 5$ and standardized y_n and x_n.

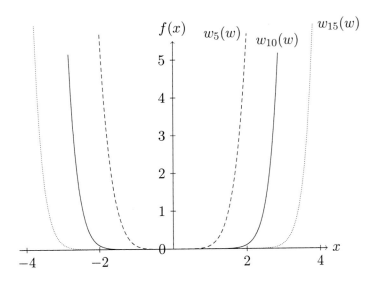

Fig. 6.3: Weight decay for use with the exponential function for $g = 5$ (dashed), $g = 10$ (line) and $g = 15$ (dotted)

Proof.

$$E(\boldsymbol{\theta}|x) = \sum_n \left(f(x_n|\boldsymbol{\theta})\right)^2 - 2f(x_n|\boldsymbol{\theta})y_n + y_n^2$$

$$= \sum_\lambda \left(\sum_\kappa \binom{\lambda}{\kappa} \tau_{\lambda-\kappa} \tau_\kappa \sum_n x_n^\lambda - 2\tau_\lambda \sum_n y_n x_n^\lambda\right) + \sum_n y_n^2$$

$$\leq E(T^{(g)}(X, y))$$

$$+ \sum_{\lambda=g+1} \left(\max_{k \leq \lambda}\{f^{(k)}(0)\} \sum_\kappa \binom{\lambda}{\kappa}^2 \tau_\lambda \sum_n x_n^\lambda - 2\tau_\lambda \sum_n y_n x_n^\lambda\right)$$

as $\max\{|x_n|\}^\lambda \sum_n y_n = 0$

$$\leq E(T^{(g)}(X, y)) + \sum_{\lambda=g+1} \left(\max_{k \leq \lambda}\{f^{(k)}(0)\} \cdot n \cdot \binom{2\lambda}{\lambda} \tau_\lambda \max\{|x_n|\}^\lambda\right)$$

with $\binom{2\lambda}{\lambda} \leq \frac{(2\lambda)^\lambda}{\lambda!}$ and $\lambda! \geq \frac{\lambda}{3}^\lambda$ for $\lambda \geq 6$

$$\leq E\big(T^{(g)}(X,y)\big) + \max_{k \leq \lambda}\{f^{(k)}(0)\} \cdot n \cdot \sum_{\lambda=g+1} \Big(\tau_\lambda(6 \cdot \max\{|x_n|\})^\lambda\Big)$$

$$\leq E\big(T^{(g)}(X,y)\big) + \max_{k \leq \lambda}\{f^{(k)}(0)\} \cdot n \cdot \sum_{\lambda=g+1} \tau_\lambda(6 \cdot \max\{|x_n|\})^\lambda$$

$\square$

Comment 6.27

The inequalities used in the proof of Theorem 6.26 are very crude. Especially, value 6 in term $(6 \cdot \max\{|x_n|\})^\lambda$ is often too high. The variable x is standardized, therefore its values are scattered around zero. This is why choosing $\max\{|x_n|\}$ is not appropriate. This value can be reduced to $\frac{\max\{|x_n|\}}{2}$ or even to $\frac{\max\{|x_n|\}}{3}$ for practical purposes.

Comment 6.28

Around zero, the weight decay penalty becomes less import for larger values of g, but it becomes more important for higher values. This is a very good feature for a weight decay penalty term which is intended to provide a soft barrier for the weights.

7 Sufficient Statistics in Multi-layer Perceptron Networks for Sequential Estimation

MLP networks are a special form of nonlinear regression functions. This is why we can use the results from the preceding chapter to construct real online learning neural networks. Taylor series expansions are used to realize online learning networks similar to general nonlinear regression and classification.

The general idea of using Taylor series expansions in MLP networks is not innovative. These series have already been used for interpretation of its weights (Beynon, Curry and Morgan, 1999) and to test the network model (Teräsvirta, Lin and Granger, 1993; Fine, 1999).

Transfer functions in MLP networks have to be non-polynomial to be able to approximate arbitrary functions (Leshno, Lin, Pinkus and Schocken, 1993). That is why we cannot expect a finite statistic to fulfill these requirements.

We have to build up power series expansions of the MLP neural network (cf. section 7.1) in order to support sufficient statistics. After looking at some numerical problems (cf. section 7.2), we are able to construct MLP networks for regression (cf. section 7.3) and (ordinal) classification (cf. section 7.4).

7.1 Taylor Series in Multi-layer Perceptron Networks

Let

$$g(a_j) = \sum_{\lambda=0}^{\infty} \kappa_\lambda a_j^\lambda$$

be the Taylor series expansion of the transfer function around zero with coefficients κ. a is given in neural networks by a sum of the type

$$a_j = w_j + \sum_{i \to j} w_{ij} x_i \ .$$

Setting $x_0 := 1$ we can use a scalar product instead of the inner sum and the Taylor series becomes (cf. Theorem A.9)

$$g(a_j) = \sum_{\lambda=0}^{\infty} \kappa_\lambda \left(\langle \boldsymbol{w_j}, \boldsymbol{x} \rangle \right)^\lambda$$

For the complete two-layered network (4.1) holds

$$f_k(\boldsymbol{X}|\boldsymbol{w}) = g_2 \left(w_k + \sum_{j \to k} w_{jk} g_1 \left(\langle \boldsymbol{w_j}, \boldsymbol{x} \rangle \right) \right)$$

$$= g_2 \left(\sum_{\lambda=0}^{\infty} \mathbb{1}_{\lambda=0}(w_k) + \sum_{k_1,\dots,k_d} \binom{\lambda}{k_1,\dots,k_d} \left(\sum_j w_{jk} \kappa_{j,\lambda} \prod_{i=1}^{d} w_{ji}^{k_i} \right) \prod_{i=1}^{d} x_i^{k_i} \right)$$

$$= \sum_{\iota=0}^{\infty} \kappa_{2,\iota} \left[w_k + \sum_{\lambda=0}^{\infty} \sum_{k_1,\dots,k_d} \binom{\lambda}{k_1,\dots,k_d} \left(\sum_j w_{jk} \kappa_{j,\lambda} \prod_{i=1}^{d} w_{ji}^{k_i} \right) \prod_{i=1}^{d} x_i^{k_i} \right]^{\iota}$$

with the binomial theorem

$$= \sum_{\iota=0}^{\infty} \kappa_{2,\iota} \sum_{\ell=0}^{\iota} \binom{\iota}{\ell} w_k^{\iota-\ell} \left[\sum_{\lambda=0}^{\infty} \sum_{k_1,\dots,k_d} \binom{\lambda}{k_1,\dots,k_d} \left(\sum_j w_{jk} \kappa_{j,\lambda} \prod_{i=1}^{d} w_{ji}^{k_i} \right) \prod_{i=1}^{d} x_i^{k_i} \right]^{\ell}$$

using Theorem A.5

$$= \sum_{\iota=0}^{\infty} \kappa_{2,\iota} \sum_{\ell=0}^{\iota} \binom{\iota}{\ell} w_k^{\iota-\ell} \sum_{\lambda=0}^{\infty} \sum_{\substack{\sum_m c_m = \lambda \\ c_m \geq 0}}$$

$$\cdot \prod_{m=1}^{\ell} \left[\sum_{\sum_{i=1}^{d} k_i = c_m} \binom{c_m}{k_1,\dots,k_d} \left(\sum_j w_{jk} \kappa_{j,\lambda} \prod_{i=1}^{d} w_{ji}^{k_i} \right) \prod_{i=1}^{d} x_i^{k_i} \right]$$

setting the index set $K_{c_m} = \{k_1 \dots k_d : \sum_{i=1}^{d} k_i = c_m\}$

$$= \sum_{\iota=0}^{\infty} \kappa_{2,\iota} \sum_{\ell=0}^{\iota} \binom{\iota}{\ell} w_k^{\iota-\ell} \sum_{\lambda=0}^{\infty} \sum_{\substack{m \\ c_m \geq 0}}^{c_m = \lambda}$$

$$\cdot \sum_{\substack{k_{i_m} \in K_{c_m} \\ m \in \{1,\dots,\ell\}}} \prod_{m=1}^{\ell} \left[\binom{c_m}{k_{1_m}, \dots, k_{d_m}} \left(\sum_j w_{jk} \kappa_{j,\lambda} \prod_{i=1}^{d} w_{ji}^{k_{i_m}} \right) \right] \prod_{i=1}^{d} x_i^{\sum_{m=1}^{\ell} k_{i_m}} .$$

$$(7.1)$$

Now the coefficients of the Taylor series expansions of transfer functions $(\kappa_1)_\lambda$ and $(\kappa_2)_\iota$ have to be identified.

7.1.1 Taylor Series Expansions of Common Transfer Functions

The two most common nonlinear sigmoidal transfer functions are the logistic function and the hyperbolic tangent (Bishop, 1995, p. 126f.).

Hyperbolic Tangent

Let g be the hyperbolic tangent. Its Taylor series expansion around 0 is

$$g(a) = \tanh(a) = \sum_{k=0}^{\infty} \frac{(-1)^k 2^{2(k+1)} \left(2^{2(k+1)} - 1 \right) B_{k+1} a^{2k+1}}{(2(k+1))!}, \quad |a| < \frac{\pi}{2}.$$

$$(7.2)$$

Logistic Function

Let g be the logistic function,

$$g(a) = \frac{1}{1 + \exp(-a)}$$

Its Taylor series expansion around 0 is

$$g(a) = \frac{1}{2} + \sum_{k=0}^{\infty} \frac{(-1)^k \left(2^{2(k+1)} - 1 \right) B_{k+1} a^{2k+1}}{(2(k+1))!}, \quad |a| < \pi \qquad (7.3)$$

Both functions are popular because their derivatives can be calculated in a

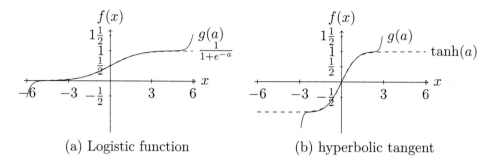

(a) Logistic function   (b) hyperbolic tangent

Fig. 7.1: Approximations of the logistic and the hyperbolic tangent function with Taylor series expansions with the first 10 summands

simple way(Bishop, 1995). This benefit is no longer required by using the set of statistics (6.17). Instead, the transfer functions should have an absolutely convergent Taylor series expansion on $\mathbb{R}$.

7.1.2 Using the Cumulative Normal Distribution as Transfer Function

The Taylor series expansions of the tanh and logistic transfer function are not absolutely convergent on $\mathbb{R}$.

Therefore, another transfer function with a simple Taylor series representation which is based on the absolutely convergent Taylor series expansion of the exponential function

$$g(a) = \frac{1}{\sqrt{2\pi}} \sum_{k=0}^{\infty} \frac{(-1)^k a^{2k+1}}{2^k (2k+1) \cdot k!} + \frac{1}{2} \tag{7.4}$$

should be used. It is the formal integral of the Taylor series expansion of the standard-normal distribution.

$$g'(a) = \frac{1}{\sqrt{2\pi}} \sum_{k=0}^{\infty} \frac{(-1)^k a^{2k}}{2^k \cdot k!}$$

$$= \frac{1}{\sqrt{2\pi}} \exp\left(-\frac{a^2}{2}\right). \tag{7.5}$$

Comment 7.1

The normal distribution has been chosen only *because of the simple structure of its Taylor series expansion. There is explicitly no other reason for it. It has been chosen because it is based on the exponential function which is absolute convergent on* $\mathbb{R}$. *Using the squared* x *is the most simple possibility of obtaining a function which tends to zero for* $|x| \to \infty$. *The integral of the function is only bounded in this case. This is necessary for a sigmoidal function. Putting it all together, the integral of* $\exp(x^2)$ *is the natural choice to construct a sigmoidal transfer function with an absolute convergent Taylor series expansion.*

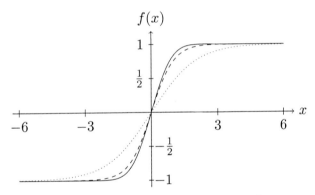

dotted: logistic, dashed: hyperbolic tangent, line: gaussian

Fig. 7.2: Transfer functions

The choice of this particular transfer function has only a slight influence on the properties of the learned parameters (see Figure 7.2). The Gaussian transfer function is (after an appropriate scaling) very close to the hyperbolic tangent and the logistic function. Figure 7.2 shows that the gradient of the Gaussian function is larger than that of the other two functions without any scaling.

There are different authors stating tanh to be "better" than the logistic transfer function as it is antisymmetric (Bishop (1995, p. 127), Haykin (1999, p. 179) and Duda, Hart and Stork (2000, p. 308)). This is why the following symmetric transfer function is used in the rest of this thesis (see Figure 7.3):

Definition 7.2 (Taylor series of the Gaussian transfer function)
The function

$$g(a) = \sqrt{\frac{2}{\pi}} \sum_{k=0}^{\infty} \frac{(-1)^k a^{2k+1}}{2^k (2k+1) \cdot k!} = \sqrt{\frac{2}{\pi}} \sum_{k=0}^{\infty} \frac{(-1)^k}{2^k (2k+1) \cdot k!} a^{2k+1} \qquad (7.6)$$

with its derivative

$$g'(a) = \sqrt{\frac{2}{\pi}} \sum_{k=0}^{\infty} \frac{(-1)^k a^{2k}}{2^k \cdot k!} \qquad (7.7)$$

is called the Gaussian transfer function.
For the sake of simpler calculation, it is rewritten to

$$g(a) = \sum_{k=0}^{\infty} \gamma_k a^k$$
$$\gamma_k = \begin{cases} \sqrt{\frac{2}{\pi}} \dfrac{(-1)^{\frac{k-1}{2}}}{2^{\frac{k-1}{2}} k \cdot (\frac{k-1}{2})!} & k \ \text{uneven} \\ 0 & \text{else} \end{cases} \qquad (7.8)$$

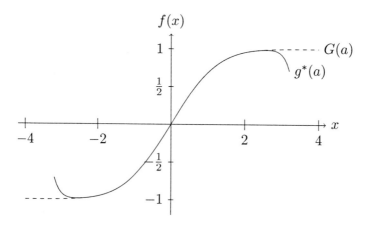

Fig. 7.3: Approximation of the Gaussian transfer function by a Taylor series
expansion with the first 10 summands

Comment 7.3

Using the normal instead of the standard normal cumulative distribution function, i.e.

$$g_\sigma^*(a) = \frac{1}{\sqrt{2\pi\sigma^2}} \sum_{k=0}^{\infty} \frac{(-1)^k a^{2k+1}}{\sigma^{2k} 2^k (2k+1) \cdot k!} + \frac{1}{2} ,$$

would introduce the redundant parameter σ to the neural network model (Thimm, Fiesler and Moerland, 1995).

However, the hyperbolic tangent and the logistic function can be "approximated" by g_σ:

$$\frac{1}{1+e^{-a}} \approx g_{1.70174453873625}^*(a)$$

$$\text{with } \|g_{1.70174453873625}^*(a) - \frac{1}{1+e^{-a}}\|^2 = 8.943995585 \cdot 10^{-5}$$

$$\tanh(a) \approx g_{0.850872314186378}(a)$$

$$\text{with } \|g_{0.850872314186378}(a) - \tanh(a)\|^2 = 3.577620932 \cdot 10^{-4}$$

Unfortunately, the convergence of the Taylor series expansion of the Gaussian transfer function is very slow. Using another sigmoidal transfer function with faster convergence would reduce the amount of necessary computation time substantially.

7.2 Numerical Contributions

From a numerical point of view, the g-approximative statistic is very problematic. The number of stored values grows exponentially with g and d. As every of these statistics has to be updated with every new observation, the calculation complexity is also exponential in g and d. This enormous computational complexity and memory usage can only be justified by improvement in accuracy, interpretability or calculation savings in other parts of the system.

7.2.1 Influence of the Statistic on the Complexity of the Neural Network Learning

With w denoting the number of parameters (weights) in a MLP, the complexity[91] of calculating and learning one pattern is $O(w)$ (Bishop, 1995). Therefore, the effort needed for online learning of n patterns is $O(n \cdot w)$. Let $o_1 \in \mathbb{N}$ denote the number of iterations through the pattern set in batch learning[92]. Then, the computational complexity is given by $O(n \cdot w \cdot o_1)$.

The number of weights w in a three-layered completely connected feed-forward neural network can be calculated with the help of the dimensions of input (d), output (k) and the number of nodes in the hidden layer (m) by $(d + k + 1) \cdot m + d_1$. Memory usage is calculated in the same way and shown in tabular 7.1 where $o_2(d, k, m)$ denotes the additional space required for the numerical optimization algorithm, e.g. for storing higher order derivatives.

Theorem 7.4
Let $X \in \mathbb{R}^d$. The needed space for the g-approximative infinite statistic $T(X|g)$ is $\binom{g+d}{d}$ and the number of basic calculations (additions and multiplications) for a sequence of $n \in \mathbb{N}$ vectors $\boldsymbol{x}_n$ is $\binom{g+d}{d} \cdot n \cdot d$ (cf. equation 6.18).

Proof. Looking at (7.1), we only need that part of the g-approximative statistic where the sum of the exponents of the different input variables X is equal or less than g. Because there are $\binom{k+d-1}{k}$ different possibilities to obtain k with d non-negative integer-valued summands, there are

$$\sum_{k=0}^{g} \binom{k+d-1}{d-1} = \sum_{k=d-1}^{g+d-1} \binom{k}{d-1} = \binom{g+d}{d}$$

possibilities for equal or less than g in sum. $\qquad \square$

Comment 7.5
Let $X \in \mathbb{R}^d$, $Y \in \mathbb{R}^k$. Let m denote the number of nodes in the hidden layer.

---

[91] For calculation with the help of the big O notation confer Graham, Knuth and Patashnik (1994).

[92] This number is stochastic due to the random initialization of the weights and the pattern set (X, y).

Then the memory usage for the MLP and its g-approximative infinite set of statistics for estimation of the parameters is

$$(d + k) \cdot m + k \cdot \binom{g + d}{d} + o_2(d, k, m)$$

and the number of calculations is

$$k \cdot \binom{g + d}{d} \cdot n + (d + k) \cdot m \cdot o_1(d, k, m) \ .$$

algorithm	computation	memory
batch	$n \cdot m \cdot (d + k) \cdot o_1(d, k, m)$	$(d + k)(m + n) + o_2(d, k, m)$
sequential	$n \cdot m \cdot (d + k)$	$m \cdot (d + k) + o_2(d, k, m)$
statistic	$k \cdot \binom{g+d}{d} \cdot n + m(d + k) \cdot o_1(d, k, m)$	$m \cdot (d + k) + k \cdot \binom{g+d}{d} + o_2(d, k, m)$

Tab. 7.1: Algorithmic efforts

The memory usage of the algorithm which is based on the g-approximative infinite set of statistics is $k \cdot \binom{g+d}{d}$ higher than that of the online learning algorithm. The effort in calculation is linear in n which is equal to that in online and offline learning (see Table 7.1). For example, iterating the gradient descent algorithm only one time $(o_1(d, k, m) = 1)$ reduces the calculation effort of the new method and so it becomes equal to online backpropagation.

Example 7.6
Let us assume that $g = 10$ is sufficient[93] and we have $d = 5$ $(d = 10)$ input dimensions and $k = 1$ output dimensions. Let us further assume the (not unusual, cf. Berry and Linoff (1997)) size of $m = 10$. Neglecting memory usage of the numerical optimization algorithm, this leads to a memory requirement of 3 063 (184 866) real values. Although this looks quite big[94], the memory usage is lower, if the number of patterns n is greater than 613 (18 487).

Because of the dominance of $\binom{g+d}{d}$, the values for g and d should be as small

---

[93]Example 7.17 shows that $g = 10$ is sufficient for the approximation of a MLP network with linear output and with approximation error $\epsilon = 0.03$.

[94]This number has to be compared to the usual memory consumption of an EBPN, not to real memory sizes.

as possible. Especially g can be approximated better as the approximation (cf. corollary 7.16) is very rough. For example, $\epsilon = 0.0074$ can be obtained for $g = 10$ by using (7.11) instead of corollary 7.16. There is another alternative to reduce the number of statistics:

Theorem 7.7

Let the two input variables X_i and X_j, $i \neq j$ be stochastic independent. Then, the minimally required number of statistics is reduced by $\frac{(g-1)g}{2}$ through setting

$$\sum_n (X_i^{(n)})^k (X_j^{(n)})^\ell = \frac{1}{n} \sum_n (X_i^{(n)})^k \cdot \sum_n (X_j^{(n)})^\ell$$

Proof. Let X_i and X_j be independent. Then the equation $\mathbb{E}X_i^k X_j^\ell = \mathbb{E}X_i^k \cdot \mathbb{E}X_j^\ell$ holds for all $k, \ell \in \mathbb{N}$. The strong law of large numbers states that $\sum_n (X_i^{(n)})^k (X_j^{(n)})^\ell \to \mathbb{E}X_i^k X_j^\ell$ and $\sum_n (X_i^{(n)})^k \overset{a.s.}{\to} \mathbb{E}X_i^k$ as $n \to \infty$. The number of statistics which can be replaced in this way is

$$\sum_{m=0}^{g-2} \binom{k+1}{1} = \sum_{m=1}^{g-1} k = \frac{(g-1)g}{2}$$

as $k, \ell \geq 1$. $\qquad\qquad\qquad\qquad\qquad\qquad\qquad\qquad\qquad\qquad\qquad\qquad\square$

Comment 7.8

The result of Theorem 7.7 can be generalized to more than two stochastically independent variables. If all variables are mutually independent, the number of needed statistics is $g \cdot d$.

Most of the attributes in large data sets contain categorical data with k_i categories that are coded with k_i or $k_i - 1$ binary variables. In this case, the number of needed statistics has to be calculated in another way than for metric variables because with $X \in \{0, 1\}$ for $j \in \mathbb{N}^{>0}$ we have $X^j = X$.

Theorem 7.9

Let X_i be categorical with k_i categories and be coded with the simple coding *scheme with a reference class (i.e. $k_i - 1$ binary variables). Let μ denote the number of metric variables. Then the total memory usage for the statistics is*

$$\sum_{m=0}^{g}\sum_{j=0}^{m}\binom{m-j+\mu-1}{\mu-1}\cdot\binom{\sum_i(k_i-1)}{j}\ .$$

Proof. The value of X is either zero or one for each dimension of a categorical variable with the simple coding scheme. Therefore, the potency is irrelevant and we only need to multiply the number of statistics necessary for the metric variables with the number of possible combinations of the k_i categories

$$\sum_{\ell=0}^{\min\{k_i,g\}}\binom{\min\{k_i,g\}}{\ell}\ . \tag{7.9}$$

$\square$

Example 7.10
Let five of the variables in example 7.6 be nominal with $k_i = 4$ for all i. Then the required memory for 15 binary and 5 metric inputs is $3\,075\,282$ values. This is equal to a corresponding data set with $153\,765$ patterns.

Comment 7.11
The result from Theorem 7.9 is only dependent on the coding scheme of the categorical variables. That is why ordinal and nominal variables can be treated in the same way. Symmetric ordinal data are a special case as they require positive and negative values of the coding. Therefore, the equation $X^j = X$ for $j \in \mathbb{N}^{>0}$ has to be replaced by $X^j = X$ for even j and $X^j = -X$ for uneven j. This means that the number of statistics concerning the ordinal variable has to be doubled. All in all, there is no difference in memory usage for the statistics to other categorical variables.

In contrast to single values, the sums within the statistics are advantageous, if a quite small g is used: The risk of creating values "out of range" is smaller in comparison to non-aggregated data because of the tendency towards the "middle". In other words, the summation reduces the influence of extremely large or small values.

7.2.2 Parallel Computation of the Statistic in Neural Networks

The next question is how to parallelize the computation and to distribute the required memory for the additional information provided by the statistic.

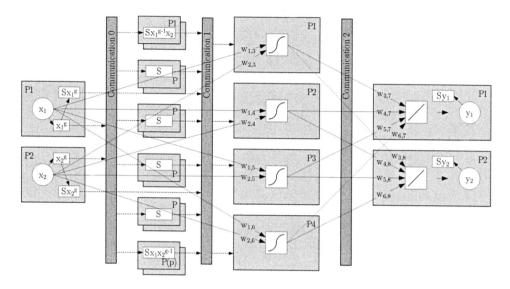

Fig. 7.4: Parallelization scheme

The proposed distribution of the statistic is illustrated in Figure 7.4 for a network with two inputs, two outputs and four hidden nodes (cf. Figure 4.4). Obviously, using sufficient statistics introduces an additional "layer" with nodes for saving the values of the statistics. The statistics, which are associated with the input and the output nodes can be calculated in parallel. However, there is an asymmetry as there are more statistics per node in the input layer than in the output layer. Furthermore the statistics of the input layer are needed for calculation of the statistics in the statistics "layer".No statistic is saved within the nodes of the hidden layer. This allows more flexibility for tuning the model complexity by changing the number of hidden nodes.

During data collection, the nodes of the input layer have to communicate with the nodes of the statistics "layer" and the output layer to calculate "mixed" statistics, i.e. the sufficient statistics consisting of more than one variable. It can be expected that the communication effort is higher than the time saving obtained by parallelization because the calculation operations are simple potentiation and addition.[95] The statistics of the statistics layer will not be calculated completely in parallel. Nevertheless, we can assume to calculate with at least as many processors in parallel, as there are input and output variables. That means that this parallelization scales approximately with factor $d + k$.

During estimation, the weight optimization has to be calculated by the output and the hidden layer and communication is required between the input layer, which provides values of the stored statistics, with the middle layer and the output layer. The derivatives for the weight updates are calculated with the help of the backpropagation algorithm in each node for the incoming weights. Here we can achieve an approximative scale factor for the parallelization of $h + k$.

7.3 Regression

MLP neural networks have to be designed with linear output nodes for regression problems. Thus, reducing the model (7.1) to

$$f(\boldsymbol{X}|\boldsymbol{w}) = w_0 + \sum_{\lambda=0}^{\infty} \sum_{k_1,\dots,k_d} \binom{\lambda}{k_1,\dots,k_d} \left(\sum_j w_j \kappa_{j,\lambda} \prod_{i=1}^{d} w_{ji}^{k_i} \right) \prod_{i=1}^{d} x_i^{k_i} \quad (7.10)$$

7.3.1 Sufficiency of the Statistic

Regarding to Theorem 6.16 and comments 6.17 and 6.21, we can state for MLP networks:

Theorem 7.12

Let (X,Y) be a random vector with elements $X \in \mathbb{R}^{d_1}$ and $Y \in \mathbb{R}$. A sequence of N i.i.d. random vectors $(X,Y)^{(n)}$ is given to fit a MLP $f(X|W^{(1)},W^{(2)})$ with transfer functions as defined by (7.6). And let T be a set of infinite statistics with $T = (T(X), T(X,Y), T_s^2(Y))$. Then the LS estimators of the parameter matrices $W^{(1)}$ and $W^{(2)}$ can only be calculated from statistic T, i.e. the statistic T is LS-sufficient.

---

[95]For example using cluster computing, where communication has to be done with the help of distributed memory or messages.

Proof. The LS estimator can be calculated by minimizing (2.5). Therefore, three different sums have to be evaluated:

$$\sum_{n=1}^{N} \left(y^{(n)}\right)^2, \quad \sum_{n=1}^{N} y^{(n)} \left(\sum_{j=0}^{M} w_j^{(2)} g \left(\sum_{i=0}^{d} w_{ji}^{(1)} x_i^{(n)}\right)\right)$$

$$\text{and} \sum_{n=1}^{N} \left(\sum_{j=0}^{M} w_j^{(2)} g \left(\sum_{i=0}^{d} w_{ji}^{(1)} x_i^{(n)}\right)\right)^2$$

where the first sum is included in T. Let τ_ℓ be the ℓ-th coefficient of the power series (7.6), i.e. defined in (7.8). Then

$$\sum_{n=1}^{N} y^{(n)} \sum_{j=0}^{M} w_{kj}^{(2)} g \left(\sum_{i=0}^{d_1} w_{ji}^{(1)} x_i^{(n)}\right)$$

$$= \sum_{j=0}^{M} w_j^{(2)} \sum_{n=1}^{N} y^{(n)} \sum_{\ell=0}^{\infty} \tau_\ell \left(\sum_{i=0}^{d_1} w_{ji}^{(1)} x_i^{(n)}\right)^\ell$$

using absolute convergence of (7.6)

$$= \sum_{j=0}^{M} w_j^{(2)} \sum_{\ell=0}^{\infty} \tau_\ell \sum_{n=1}^{N} y^{(n)} \left(\sum_{i=0}^{d_1} w_{ji}^{(1)} x_i^{(n)}\right)^\ell$$

and with the multinomial theorem we get

$$= \sum_{j=0}^{M} w_j^{(2)} \sum_{\ell=0}^{\infty} \tau_\ell \sum_{n=1}^{N} y^{(n)} \sum_{\kappa_0,\dots,\kappa_{d_1}} \binom{\ell}{\kappa_0,\dots,\kappa_{d_1}} \prod_{i=0}^{d_1} \left(w_{ji}^{(1)} x_i^{(n)}\right)^{\kappa_i}$$

$$= \sum_{j=0}^{M} w_j^{(2)} \sum_{\ell=0}^{\infty} \tau_\ell \sum_{\kappa_0,\dots,\kappa_{d_1}} \binom{\ell}{\kappa_0,\dots,\kappa_{d_1}} \prod_{i=0}^{d_1} \left(w_{ji}^{(1)}\right)^{\kappa_i} \sum_{n=1}^{N} y^{(n)} \prod_{i=0}^{d_1} \left(x_i^{(n)}\right)^{\kappa_i}$$

$\sum_{n=1}^{N} y^{(n)} \prod_{i=0}^{d_1} \left(x_i^{(n)}\right)^{\kappa_i}$ is included in $T(X,Y)$ for all i. With an analogous argumentation it is shown that the third sum can be calculated from $T(X)$:

$$\sum_{n=1}^{N} \left(\sum_{j=0}^{M} w_j^{(2)} g\left(\sum_{i=0}^{d} w_{ji}^{(1)} x_i^{(n)} \right) \right)^2$$

using the scalar product and Theorem A.9

$$= \sum_{n=1}^{N} \left(w_0 + \sum_{\lambda=0}^{\infty} \sum_{k_0,\dots,k_d} \binom{\lambda}{k_0,\dots,k_d} \left(\sum_j w_j \kappa_{j,\lambda} \prod_{i=1}^{d} w_{ji}^{k_i} \right) \prod_{i=1}^{d} x_i^{k_i} \right)^2$$

again, with absolute convergence

$$= N w_0^2 + 2 w_0 \sum_{\lambda=0}^{\infty} \sum_{k_0,\dots,k_d} \binom{\lambda}{k_0,\dots,k_d} \left(\sum_j w_j \kappa_{j,\lambda} \prod_{i=1}^{d} w_{ji}^{k_i} \right) \sum_{n=1}^{N} \prod_{i=1}^{d} x_i^{k_i}$$

$$+ \sum_{n=1}^{N} \left(\sum_{\lambda=0}^{\infty} \sum_{k_0,\dots,k_d} \binom{\lambda}{k_0,\dots,k_d} \left(\sum_j w_j \kappa_{j,\lambda} \prod_{i=1}^{d} w_{ji}^{k_i} \right) \prod_{i=1}^{d} x_i^{k_i} \right)^2$$

We concentrate on the last summand, and by using Theorem A.5 we obtain:

$$\sum_{n=1}^{N} \left(\sum_{\lambda=0}^{\infty} \sum_{k_0,\dots,k_d} \binom{\lambda}{k_0,\dots,k_d} \left(\sum_j w_j \kappa_{j,\lambda} \prod_{i=1}^{d} w_{ji}^{k_i} \right) \prod_{i=0}^{d} x_i^{k_i} \right)^2$$

$$= \sum_{n=1}^{N} \sum_{\lambda=0}^{\infty} \sum_{\ell_1+\ell_2=\lambda} \left(\sum_{k_0,\dots,k_d} \binom{\ell_1}{k_0,\dots,k_d} \left(\sum_j w_j \kappa_{j,\ell_1} \prod_{i=0}^{d} w_{ji}^{k_i} \right) \prod_{i=0}^{d} x_i^{k_i} \right)$$

$$\cdot \left(\sum_{k_0,\dots,k_d} \binom{\ell_2}{k_0,\dots,k_d} \left(\sum_j w_j \kappa_{j,\ell_2} \prod_{i=0}^{d} w_{ji}^{k_i} \right) \prod_{i=0}^{d} x_i^{k_i} \right)$$

$$= \sum_{n=1}^{N} \sum_{\lambda=0}^{\infty} \sum_{\ell_1+\ell_2=\lambda} \sum_{k_0,\dots,k_d} \sum_{c_0,\dots,c_d} \binom{\ell_1}{k_0,\dots,k_d} \binom{\ell_2}{c_0,\dots,c_d}$$

$$\cdot \left(\sum_j w_j \kappa_{j,\ell_1} \prod_{i=1}^{d} w_{ji}^{k_i} \right) \prod_{i=0}^{d} x_i^{k_i} \left(\sum_j w_j \kappa_{j,\ell_2} \prod_{i=0}^{d} w_{ji}^{c_i} \right) \prod_{i=0}^{d} x_i^{c_i}$$

and by considering absolute convergence of the series

$$= \sum_{\lambda=0}^{\infty} \sum_{\ell_1+\ell_2=\lambda} \sum_{k_0,\ldots,k_d} \sum_{c_0,\ldots,c_d} \binom{\ell_1}{k_0,\ldots,k_d} \binom{\ell_2}{c_0,\ldots,c_d}$$

$$\cdot \left(\sum_j w_j \kappa_{j,\ell_1} \prod_{i=0}^{d} w_{ji}^{k_i} \right) \left(\sum_j w_j \kappa_{j,\ell_2} \prod_{i=0}^{d} w_{ji}^{c_i} \right) \sum_{n=1}^{N} \prod_{i=0}^{d} x_i^{k_i+c_i}$$

□

Obviously, it is possible to use online learning by invoking a numerical minimization algorithm after each update of the set of infinite statistics.

7.3.2 Approximation

In traditional MLP networks, the commonly used transfer functions (logistic function and hyperbolic tangent) are based on the exponential function. Unfortunately, the exponential function is defined as a limiting process (sequence or series) and has no finite form. Therefore, these functions have to be approximated (Anguita, Parodi and Zunino, 1993). Using the infinite statistics, this approximation is performed more explicitly and it is tailored for the actual problem. Fortunately, the error made in limiting the approximation of the exponential series by restricting the statistics to the first g summands can be calculated depending on $W^{(1)}$, $W^{(2)}$ and the family of maximum values $\{\max_n\{|x_i^{(n)}|\}\}_i$.

First, we have to find bounds for the error made by restricting (7.6) to its first g summands.

Proposition 7.13

$$g(x) = \sqrt{\frac{2}{\pi}} \left(\sum_{k=0}^{g} \frac{(-1)^k x^{2k+1}}{2^k(2k+1)\cdot k!} + R_{g+1}(x) \right) ,$$

where

$$|R_{g+1}(x)| \leq \frac{|x|^{2g+3}}{2^g(2g+3)(g+1)!} \quad \forall_x |x| \leq \sqrt{g+2} . \tag{7.11}$$

Proof.

$$|R_{g+1}(x)| \leq \sum_{k=g+1}^{\infty} \frac{|x|^{2k+1}}{2^k(2k+1) \cdot k!}$$

$$\leq \frac{|x|^{2g+3}}{2^{g+1}(2g+3)(g+1)!}$$

$$\left[1 + \frac{x^2}{2(g+2)} + \left(\frac{x^2}{2(g+2)}\right)^2 + \cdots\right]$$

for $|x| \leq \sqrt{g+2}$, we have a geometric series with $1 + \frac{x^2}{2(g+2)} + \cdots \leq 2$. □

Using this result, we can control the approximation error of the error function.

Theorem 7.14
The error of approximating (4.1) is limited for
$\max_{i,j} |\sum_{i=0}^{d_1} w_{ji}^{(1)} x_i^{(n)}| \leq \sqrt{g+2}$ *by*

$$\frac{\sqrt{g+2}^{2g+3}}{2^g(2g+3)(g+1)!} \ . \tag{7.12}$$

Proof.

$$\max \left| \sum_{j=0}^{M} w_j^{(2)} \tilde{g}\left(\sum_{i=0}^{d_1} w_{ji}^{(1)} x_i\right) - \sum_{j=0}^{M} w_{kj}^{(2)} g\left(\sum_{i=0}^{d_1} w_{ji}^{(1)} x_i\right) \right|$$

using (7.6), with τ_ℓ denoting the coefficients

$$\leq \max \sum_{j=0}^{M} \left| w_{kj}^{(2)} \left(\tilde{g}\left(\sum_{i=0}^{d_1} w_{ji}^{(1)} x_i\right) - g\left(\sum_{i=0}^{d_1} w_{ji}^{(1)} x_i\right)\right) \right|$$

and with lemma 7.13 for $|\sum_{i=0}^{d_1} w_{ji}^{(1)} x_i| \leq \sqrt{g+2}$

$$\leq \max \sum_{j=0}^{M} \left| w_j^{(2)} \left(\frac{|\sum_{i=0}^{d_1} w_{ji}^{(1)} x_i|^{2g+3}}{2^g(2g+3)(g+1)!}\right) \right| \ .$$

□

Comment 7.15

Because of the exponential dependence on $\max_i |x_i^{(n)}|$ huge values for any x_i have to be avoided and standardized inputs should be used.

Today, MLPs networks are usually implemented on digital (desktop) computers. The precision of calculation of the exponential function is restricted by the length of the digital number representation. To avoid overflows in a calculation with logistic function the bounds

$$\dot{g}(x) = \begin{cases} 0 & x \leq -45 \\ g(x) & -45 < x < 45 \\ 1 & x \geq 45 \end{cases} \tag{7.13}$$

are proposed and should work with double-precision machines (Sarle, 1997b). This is the reason why these bounds have to be obeyed in actual implementations of neural networks. We should calculate the minimum value for g, such that we are able to approximate this restricted function below a certain maximum error.

Corollary 7.16

Let ϵ be the maximum error which is accepted in approximating $g(\cdot)$. With $g \geq \sqrt[3]{2} \cdot \frac{3}{2} \cdot z_{1-\frac{\epsilon}{2}}^2 \cdot (\sqrt[6]{\epsilon})^{-1}$, it holds

$$|\tilde{g}(x) - g(x)| \leq \epsilon \quad \forall x \in \mathbb{R}$$

when

$$\tilde{g}(x) = \begin{cases} 0 & x \leq -z_{1-\frac{\epsilon}{2}} \\ g(x) & -z_{1-\frac{\epsilon}{2}} < x < z_{1-\frac{\epsilon}{2}} \\ 1 & x \geq z_{1-\frac{\epsilon}{2}} \end{cases} .$$

Proof. Because of the symmetry of $g(\cdot)$, the borders—like in (7.13)—can be obtained from

$$|1 - (2 \cdot \Phi(x) - 1)| \leq \epsilon \Leftrightarrow |x| \geq z_{1-\frac{\epsilon}{2}} .$$

Thus, $g \geq (z_{1-\frac{\epsilon}{2}})^2 - 2$, with Lemma 7.13 and $g \geq 4$ we get:

$$\frac{|z_{1-\frac{\epsilon}{2}}|^{2g+3}}{2^g (g+2)!} \leq \frac{4 \cdot (\frac{z_{1-\frac{\epsilon}{2}}^2}{2})^{g+2}}{(\frac{g+2}{3})^{g+2}} \leq 4 \cdot \left(\frac{3 \cdot z_{1-\frac{\epsilon}{2}}^2}{2 \cdot (g+2)} \right)^6 \leq \epsilon$$

□

Example 7.17

Let us assume $\epsilon = 0.03$ would be sufficient for some purposes. Then we get $g \geq 10$ (cf. corollary 7.16).

7.3.3 Extended Example for Multi-layer Perceptron Network Regression

Starting with equation (7.10), we have to calculate three distinct sums. More specifically, let us have a neural network with one metric input($d = 1$), two nodes in the hidden layer ($M = 2$) and one metric output ($k = 1$). With the help of the LS-sufficient set of infinite statistics we are able to calculate the two values using the transfer function (7.6) for both hidden nodes:

$$\sum_{n=1}^{N} y^{(n)} \sum_{j=0}^{M} w_j^{(2)} g\left(\sum_{i=0}^{d_1} w_{ji}^{(1)} x_i^{(n)}\right)$$

$$= w_0 \left[\sum_{n=1}^{N} y^{(n)}\right] + \sum_{j=1}^{2} w_j \sum_{\lambda=0}^{\infty} \tau_\lambda \sum_{\kappa_0, \kappa_1} \binom{2\lambda + 1}{\kappa_0, \kappa_1} w_{j0}^{\kappa_0} w_{j1}^{\kappa_1} \left[\sum_{n=1}^{N} y^{(n)} \left(x_1^{(n)}\right)^{\kappa_1}\right]$$

and

$$\sum_{n=1}^{N} \left(\sum_{j=0}^{2} w_j^{(2)} g\left(\sum_{i=0}^{d} w_{ji}^{(1)} x_i^{(n)}\right)\right)^2$$

$$= N w_0^2 + 2 w_0 \sum_{\lambda=0}^{\infty} \sum_{k_0, k_1} \binom{2\lambda + 1}{k_0, k_1} \tau_\lambda \left(\sum_{j=1}^{2} w_j \prod_{i=0}^{1} w_{ji}^{k_i}\right) \left[\sum_{n=1}^{N} x_i^{k_i}\right]$$

$$+ \sum_{\lambda=0}^{\infty} \sum_{\ell_1+\ell_2=2\lambda+1} \sum_{k_0, k_1} \sum_{c_0, c_1} \binom{\ell_1}{k_0, k_1} \binom{\ell_2}{c_0, c_1}$$

$$\cdot \tau_{\ell_1} \tau_{\ell_2} \left(\sum_{j=1}^{2} w_j \prod_{i=0}^{1} w_{ji}^{k_i}\right) \left(\sum_{j=1}^{2} w_j \prod_{i=0}^{1} w_{ji}^{c_i}\right) \left[\sum_{n=1}^{N} x_1^{k_1+c_1}\right]$$

from Theorem 7.12.

Again, we have simulated 200 data points from the function represented by a neural network with two hidden nodes in the range $[-1, 1]$ with additive

Gaussian error with zero mean and a standard deviation of $\frac{1}{5}$. These data have been used to fit the neural network with and without the help of the approximative sufficient statistics (see Figure 7.5). Using the 20th-approximative statistic, a good approximation of the data can be obtained. Furthermore, the error function is very simple compared to the error function which results from direct estimation. This indicates, that using the statistic provides a much more better estimation of the function, because the error is based on the restricted complexity of the regression function. This is why we can improve this result by using a higher degree than 20 within the approximation.

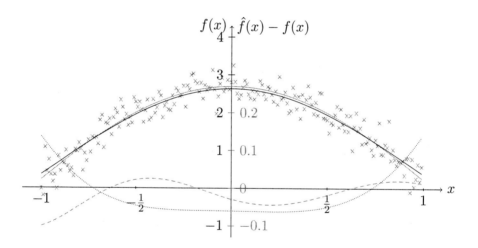

gray line: function, crosses: simulated data
black line: estimated function based on 20th-approx. statistic
dashed gray: error of direct estimation (without statistic)
dotted: error of estimation based on 20th approx. statistic

Fig. 7.5: Graphs of the function and its approximations

The data have been standardized to obtain better estimators. The following weight decay has been used (cf. Theorems 7.14 and 6.26):

$$\frac{n \cdot \sum_i |2w_i|^{g+1}}{2^{\lceil \frac{g+1}{2} \rceil}(g+1)(\lceil \frac{g+1}{2} \rceil)!}$$

7.4 Classification

MLP networks for univariate regression and those for univariate classification differ in two major details: First, the number of output nodes of a MLP network for regression is 1 and for classification is k. Secondly, the transfer functions in the nodes of the output layer are sigmoidal instead of linear for classification.

Unfortunately, there exists no on $\mathbb{R}$ convergent Taylor series expansion for the function

$$t(x) = \frac{1}{1 + e^x}.$$

Its convergence radius is just $(-\pi, \pi)$ (see Figure 7.6).

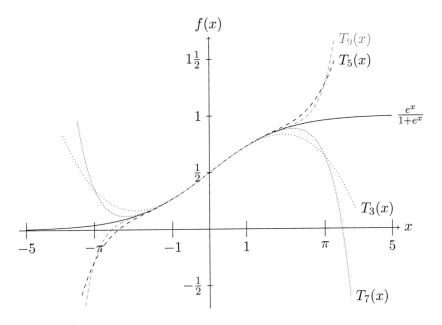

Fig. 7.6: Taylor series approximations to the natural link function

This problem could be solved by using another CDF, for example the Gaussian transfer function (7.6).[96]

The error function for classification is given by the multinomial distribution and the ML procedure can be used for estimation. Therefore, the concept of LS-sufficiency is superfluous in classification at first sight.

---

[96]The special form of the link function is irrelevant in nonlinear regression (Cox, 1984).

Sufficiency of the Statistic

The multinomial distribution belongs to the exponential family and its probability function (for a single experiment) has got the form

$$f(\boldsymbol{y}^{(n)}|\boldsymbol{\theta}) = \exp\left[\sum_{j=1}^{k} \ln\left[\frac{\pi_j}{1 - \sum_{i=1}^{k}\pi_i}\right] y_j^{(n)} - \ln\left[1 - \sum_{i=1}^{k}\pi_i\right]\right]$$

$$= \exp\left[\sum_{j=1}^{k} y_j^{(n)} f_j(\boldsymbol{x}^{(n)}|\boldsymbol{w}) - \ln\left[\frac{1}{1 + \sum_{j=1}^{k-1}\exp\left[f_j(\boldsymbol{x}^{(n)}|\boldsymbol{w})\right]}\right]\right]$$

where $f_j(\boldsymbol{x}^{(n)}|\boldsymbol{w})$ describes the $k-1$ continuous predictor variables.

Using this, the log-likelihood can be calculated:

$$\log \mathcal{L}(\boldsymbol{X}|\boldsymbol{w}) = \sum_{i=1}^{n} \log\left[\sum_{j=1}^{k-1} y_j^{(n)} f_j(\boldsymbol{x}^{(n)}|\boldsymbol{w}) - \ln\left[\frac{1}{1 + \sum_{j=1}^{k-1}\exp\left[f_j(\boldsymbol{x}^{(n)}|\boldsymbol{w})\right]}\right]\right]$$

$$= \left[\sum_{i=1}^{n}\sum_{j=1}^{k} y_j^{(n)} f_j(\boldsymbol{x}^{(n)}|\boldsymbol{w}) + \sum_{i=1}^{n} \ln\left[1 + \sum_{j=1}^{k-1}\exp\left[f_j(\boldsymbol{x}^{(n)}|\boldsymbol{w})\right]\right]\right]$$

Now, we have to have a closer look at:

$$\ln\left[1 + \sum_{j=1}^{k}\exp\left[f(x_j^{(n)}|w)\right]\right] = \sum_{\ell=1}^{\infty}\frac{1}{\ell}\left(1 - \frac{1}{1 + \sum_{j=1}^{k-1}\exp\left[f(x_j^{(n)}|w)\right]}\right)^{\ell}$$

$$= \sum_{\ell=1}^{\infty}\frac{1}{\ell}\sum_{m=0}^{\ell}\binom{\ell}{m}(-1)^m\sum_{o=0}^{\infty}\binom{-m}{o}\left(\sum_{j=1}^{k-1}\exp\left[f(x_j^{(n)}|w)\right]\right)^{-(m+o)}$$

We have to construct a power series for the logarithm which is only convergent on $(-1, 1)$. This is the reason why we cannot construct sufficient statistics. Even if we try to use another error distribution, there will always be the problem of standardizing the continuous predictor variables which implies using a fractional structure without having a convergent Taylor series expansion on whole $\mathbb{R}$.

Two possibilities for building a MLP network for categorical regression remain: First, we can use the MSE approach with LS estimation (cf. section 4.5). Secondly, we can make sure that the predictor values are restricted to the interval $(-1, 1)$ which is not very appealing.

The MSE approach is a simple application of equation (7.1). And the following theorem is obvious:

Theorem 7.18
Let (X, Y) be a random vector with elements $X \in \mathbb{R}^{d_1}$ and $Y \in \{0, 1\}^k$ and $(X, Y)^{(n)}$ be a sequence of N i.i.d. random vectors. Let $f(X|W^{(1)}, W^{(2)})$ be a MLP with transfer functions as defined by (7.6). And let T be a set of infinite statistics with $T = (T(X), T(X, Y_1 = 1), \dots, T(X, Y_{k-1} = 1), T_s^2(Y))$. Then the LS estimators to the parameter matrices $W^{(1)}$ and $W^{(2)}$ can only be obtained from the statistic T, i.e. the statistic T is LS-sufficient.

Proof. The proof is straight forward and very similar to the proof of Theorem 7.12. □

The second idea is a little bit more complicated and there is no solution for general classification. This can easily been seen from the fact that $\sum_j \exp\left[f(x_j^{(n)}|w)\right]$ has to be smaller than 1, which implies that every $f(x_j^{(n)}|w)$ has to be negative. In general, there is no technique which could provide this restriction. Nevertheless, we can expect the categories to be ordered in the predictor values for ordinal classification. Therefore—with appropriate constraints—it is possible to take this algorithmic restriction into account.

However, we may use another fix point for the Taylor series expansion of the logarithm. It is not possible to obtain a Taylor series expansion, which is convergent on the complete positive real line, but we can extend the interval of convergence to a larger region. The interval has to be large enough that the restrictions are not exploited for practical purposes.

For the moment, the first model is practicable and sufficient. Nevertheless, further research is needed to allow the use of ML estimation for MLP classification based on sufficient statistics.

8 Conclusion

Two main topics have been discussed in this thesis, both closely related to the analysis of data from large databases. At first, measurements of scales have been revisited. Most of the data stored in private and governmental databases is ordinal. These require special methods and models for their analysis in order to render as much interpretable information as possible.

Different types of ordinal data have been identified and are set into relation by a taxonomy. This taxonomy has been used to classify the various methods for ordinal data analysis known today. With the help of this taxonomy, data analysts are able to build more appropriate models for the special type of data they want to analyze.

Answering questions to the data often incorporates huge model constructions with a lot of differently scaled variables. If metric variables are included in the model, linear models are usually inappropriate. In these cases, MLP networks, which are nonlinear input-output models, can be used. It is shown in this thesis that MLP networks are an easy-to-use tool for modeling variables with arbitrary scales.[97]

Today, every data mining method should work in an online setting. This is not only important for real-time applications, but also for processing large amounts of data. Online-learnable models just require a single access to the database. They extract the required information for model estimation instantly. MLP networks have been designed for online estimation, but they either require very large samples with a lot of near-redundant observations or multiple iterations over certain sub-samples. The property of MLP networks of being online learnable is based on stochastic approximation algorithms. Stochastic estimation procedures lead to estimators that are dependent on the ordering of the observations in the sample. It is very difficult to interpret both these estimators and the resulting predictions for new observations. Furthermore,

---

[97]MLP networks can easily be adopted to every measurement of scale in input and output variables. Other aspects, especially choosing the right model size, are not trivial at all.

stochastic approximation algorithms provide estimators that suffer from high variances.

In this thesis a new estimation technique for nonlinear regression and especially for MLP networks is introduced. It is based on sequential sufficient statistics. The resulting methodology is characterized by an estimation procedure that is almost independent from data reduction. This allows data analysts to adopt the model much faster, to use MLP networks in real-time applications and to avoid parameters with questionable interpretation.

The MLP networks, which are based on sequential sufficient statistics, can be used for online estimation, online model size determination and online tracking of time-dependent relationships. These networks have to be represented by Taylor series expansions. This implies high memory usage and certain difficulties with classification problems. Parallelization and careful modeling are necessary to overcome the first shortcoming and have been discussed in this thesis. Classification based on sequentially sufficient statistics should be taken into account in future research.

This thesis provides answers to some of the most challenging problems of actual data analysis. This has been achieved by unifying aspects of algorithmic theory, statistical theory and abstract measurement theory. Particularly, cooperation of algorithmic and statistical theory is essential to solve the problems of data analysis in the future: Analysis of large databases for decision support and intelligent control of complex technical systems. Statistically well-defined models, estimated by online algorithms based on sequential sufficient statistics, provide a feasible solution to analyze large amounts of differently scaled data.

Bibliography

Agresti, A. (1984) *Analysis of Ordinal Categorical Data.* Wiley Series in Probability and Mathematical Statistics. New York: John Wiley & Sons.

Agresti, A. (1986) Applying r^2-type measures to ordered categorical data. *Technometrics*, **28**, 133–138.

Agresti, A. (1999) Modelling ordered categorical data: Recent advances and future challenges. *Statistics in Medicine*, **18**, 2191–2207.

Agresti, A. (2002) *Categorical Data Analysis.* Wiley Series in Probability and Statistics. New York: John Wiley & Sons, 2 edn.

Agresti, A., Chuang, C. and Kezouh, A. (1987) Order-restricted score parameters in association models for contingency tables. *Journal of the American Statistical Association*, **82**, 619–623.

Agresti, A. and Lang, J. B. (1993) A proportional odds model with subject-specific effects for repeated ordered categorical responses. *Biometrika*, **80**, 527–534.

Agresti, A. and Yang, M. C. (1987) An empirical investigation of some effects of sparseness in contingency tables. *Computational Statistics & Data Analysis*, **5**, 9–21.

Albert, P. S., Hunsberger, S. A. and Biro, F. M. (1997) Modeling repeated measures with monotonic ordinal responses and misclassification, with applications to studying maturation. *Journal of the American Statistical Association*, **92**, 1304–1311.

Aleksander, I. and Morton, H. (1990) *An Introduction to Neural Computing.* Chapman and Hall.

Amari, S.-i. (1990) Mathematical foundations of neurocomputing. *Proceedings of the IEEE*, **78**, 1443–1463.

Ananth, C. V. and Kleinbaum, D. G. (1997) Regression models for ordinal responses: A review of methods and applications. *International Journal of Epidemiology*, **26**, 1323–1333.

Anders, U. (1997) *Statistische Neuronale Netze.* Vahlen.

Anderson, J. A. (1984) Regression and ordered categorical variables. *Journal of the Royal Statistical Society: Series B (Methodological)*, **46**, 1–30.

Anderson, J. A. and Philips, P. R. (1981) Regression, discrimination and measurement models for ordered categorical variables. *Applied Statistics*, **30**, 22–31.

Anderson, N. H. (1961) Scales and statistics: parametric and nonparametric. *Psychological Bulletin*, **58**, 305–16.

Anguita, D., Parodi, D. and Zunino, R. (1993) Speed improvement of the back-propagation on current generation workstations. *World Congress on Neural Networking*, **1**, 165–168.

Armstrong, B. G. and Sloan, M. (1989) Ordinal regression models for epidemiologic data. *American Journal of Epidemiology*, **129**, 191–204.

Baker, B. O., Hardyck, C. D. and Petrinovich, L. F. (1966) Weak measurements vs. strong statistics: An empirical critique of S. S. Stevens' proscriptions on statistics. *Educational and Psychological Measurement*, **26**, 291–309.

Baker, R. J. and Nelder, J. A. (1978) *The GLIM System Release 3: Generalised Linear Interactive Modelling*. Numerical Algorithms Group.

Barnhart, H. X. and Sampson, A. R. (1994) Overview of multinomial models for ordinal data. *Communications in Statistics—Theory and Methods*, **23**, 3395–3416.

Barron, A. R. (1993) Universal approximation bounds for superpositions of a sigmoidal function. *IEEE Transactions on Information Theory*, **39**, 930–945.

Barron, A. R. (1994) Approximation and estimation bounds for artificial neural networks. *Machine Learning*, **14**, 115–133.

Baum, E. B. (1988) On the capabilities of multilayer perceptrons. *Journal of Complexity*, **4**, 193–215.

Bazaraa, M. S., Sherali, H. D. and Shetty, C. M. (1993) *Nonlinear Programming: Theory and Algorithms*. New York: John Wiley & Sons, 2 edn.

Behera, L., Kumar, S. and Patnaik, A. (2006) On adaptive learning rate that guarantees convergence in feedforward networks. *IEEE Transactions on Neural Networks*, **17**, 1116–1125.

Bender, R. and Benner, A. (2000) Calculating ordinal regression models in SAS and S-Plus. *Biometrical Journal*, **42**, 677–699.

Berry, M. J. A. and Linoff, G. (1997) *Data Mining Techniques*. New York: John Wiley & Sons.

Beynon, M., Curry, B. and Morgan, P. (1999) Neural networks and finite-order approximations. *IMA Journal of Management Mathematics*, **10**, 225–244.

Bishop, C. M. (1991) A fast procedure for retraining the multilayer perceptron. *International Journal of Neural Systems*, **2**, 229–236.

Bishop, C. M. (1995) *Neural Networks for Pattern Recognition*. Oxford: Oxford University Press.

Blum, J. R. (1954) Multidimensional stochastic approximation methods. *The Annals of Mathematical Statistics*, **25**, 737–744.

Boes, S. and Winkelmann, R. (2006) Ordered response models. *Allgemeines Statistisches Archiv*, **90**, 165–180.

Boyle, R. (1970) Path analysis and ordinal data. *The American Journal of Sociology*, **75**, 461–480.

Brant, R. (1990) Assessing proportionality in the proportional odds model for ordinal logistic regression. *Biometrics*, **46**, 1171–1178.

Breiman, L. (2001) Statistical modeling: The two cultures. *Statistical Science*, **16**, 199–231.

Breiman, L., Friedman, J. H., Olshen, R. A. and Stone, C. J. (1984) *Classification and Regression Trees*. New York: Chapman & Hall.

Breslaw, J. A. and McIntosh, J. (1998) Simulated latent variable estimation of models with ordered categorical data. *Journal of Econometrics*, **87**, 25–47.

Brinker, K., Fürnkranz, J. and Hüllermeier, E. (2006) Label ranking by learning pairwise preferences. Tech. Rep. TUD-KE-2007-01, Knowledge Engineering Group, TU Darmstadt. URL http://www.ke.informatik.tu-darmstadt.de/publications/reports/tud-ke-2007-01.pdf.

Brown, M., An, P. C., Harris, C. J. and Wang, H. (1993) How biased is your multi-layer perceptron? In *World Congress on Neural Networks*, vol. 3, 507–511. Portland.

Burke, C. J. (1953) Additive scales and statistics. *Psychological Review*, **60**, 73–5.

Campbell, M. K. and Donner, A. (1989) Classification efficiency of multinomial logistic regression relative to ordinal logistic regression. *Journal of the American Statistical Association*, **84**, 587–591.

Campbell, M. K., Donner, A. and Webster, K. M. (1991) Are ordinal models useful for classification? *Statistics in Medicine*, **10**, 383–394.

Cao-Van, K. and De Baets, B. (2003) Growing decision trees in an ordinal setting. *International Journal of Intelligent Systems*, **18**, 733–750.

Cardoso, J. S. and da Costa, J. F. P. (2007) Learning to classify ordinal data: The data replication method. *Journal of Machine Learning Research*, **8**, 1393–1429.

Castro, J. L., Mantas, C. J. and Benítez, J. M. (2000) Neural networks with a continuous squashing function in the output are universal approximators. *Neural Networks*, **13**, 561–563.

Cheng, B. and Titterington, D. M. (1994) Neural networks: A review from a statistical perspective. *Statistical Science*, **9**, 2–30.

Cheng, J. (2007) A neural network approach to ordinal regression. URL http://arxiv.org/abs/0704.1028.

Chu, W. and Keerthi, S. S. (2005) New approaches to support vector ordinal regression. In *Machine Learning: 16th European Conference on Machine Learning, Porto, Portugal, October 3-7, 2005. Proceedings*, vol. 119 of *ACM International Conference Proceeding Series*, 145–152. Bonn: ACM.

Chung, K. L. (1954) On a stochastic approximation method. *The Annals of Mathematical Statistics*, **25**, 463–483.

Cliff, N. (1992) Abstract measurement theory and the revolution that never happened. *Psychological Science*, **3**, 186–190.

Cliff, N. (1996a) Answering ordinal questions with ordinal data using ordinal statistics. *Multivariate Behavioral Research*, **31**, 331–350.

Cliff, N. (1996b) *Ordinal Methods for Behavioral Data Analysis*. Mahwah: Lawrence Erlbaum.

Cochran, W. G. (1977) *Sampling Techniques*. Wiley series in probability and mathematical statistics : Applied probability and statistics. New York: John Wiley & Sons, 3 edn.

Cohn, D. L. (1980) *Measure Theory*. Boston: Birkhäuser.

Cohn, H., Kleinberg, R., Szegedy, B. and Umans, C. (2005) Group-theoretic algorithms for matrix multiplication. In *Foundations of Computer Science. 46th Annual IEEE Symposium on*, 379–388.

Coppersmith, D. and Winograd, S. (1987) Matrix multiplication via arithmetic progressions. *Proceedings of the nineteenth annual ACM conference on Theory of computing*, 1–6.

Cormen, T. H., Leiserson, C. E. and Rivest, R. L. (1990) *Introduction to Algorithms*. Cambridge: MIT Press.

Couvreur, C. and Couvreur, P. (1997) Neural networks and statistics: a naïve comparison. *Belgian Journal of Operations Research, Statistics, and Computer Science*, **36**, 217–225.

Cox, C. (1984) Generalized linear models-the missing link. *Applied Statistics*, **33**, 18–24.

Cox, C. (1988) Multinomial regression models based on continuation ratios. *Statistics in Medicine*, **7**, 435–441.

Cox, C. (1995) Location-scale cumulative odds models for ordinal data: a generalized non-linear model approach. *Statistics in Medicine*, **14**, 1191–1203.

Cristianini, N. and Shawe-Taylor, J. (2000) *Support Vector Machines.* Cambridge: Cambridge University Press.

Cybenco, G. (1989) Approximation by superpositions of a sigmoidal function. *Mathematics of Control, Signals and Systems*, **2**, 303–314.

da Costa, J. F. P. and Cardoso, J. S. (2005) Classification of ordinal data using neural networks. In *Proceedings of the 16th European Conference on Machine Learning* (eds. J. Gama, R. Camacho, P. B. Brazdil, A. M. Jorge and L. Torgo), vol. 3720 of *Lecture Notes in Computer Science*, 690–697. Springer.

Daqi, G. and Genxing, Y. (2003) Influences of variable scales and activation functions on the performances of multilayer feedforward neural networks. *Pattern Recognition*, **36**, 869–878.

Dempster, A. P. and Rubin, D. B. (1983) Rounding error in regression: The appropriateness of Sheppard's corrections. *Journal of the Royal Statistical Society: Series B (Methodological)*, **45**, 51–59.

Domowitz, I. and White, H. (1982) Misspecified models with dependent observations. *Journal of Econometrics*, **20**, 35–58.

Duda, R. O., Hart, P. E. and Stork, D. G. (2000) *Pattern Classification.* New York: John Wiley and Sons.

Duin, R. P. W. (1996) A note on comparing classifiers. *Pattern Recognition Letters*, **17**, 529–536.

Efron, B. (1975) The efficiency of logistic regression compared to normal discriminant analysis. *Journal of the American Statistical Association*, **70**, 892–898.

Everitt, B. S. (1988) A finite mixture model for the clustering of mixed-mode data. *Statistics and Probability Letters*, **6**, 305–309. Besorgen? Mathe/Statistik.

Everitt, B. S. and Merette, C. (1990) The clustering of mixed-mode data: a comparison of possible approaches. *Journal of Applied Statistics*, **17**, 283–297.

Farewell, V. T. (1982) A note on regression analysis of ordinal data with variability of classification. *Biometrika*, **69**, 533–538.

Fielding, A. (1997) On scoring ordered classifications. *British Journal of Mathematical and Statistical Psychology*, **50**, 285–307.

Fielding, A. (1999) Why use arbitrary points scores?: ordered categories in models of educational progress. *Journal of the Royal Statistical Society: Series A (Statistics in Society)*, **162**, 303–328.

Fine, T. L. (1999) *Feedforward Neural Network Methodology*. Springer.

Frank, E. and Hall, M. (2001) A simple approach to ordinal classification. In *Machine Learning: 12th European Conference on Machine Learning*, vol. 2167 of *Lecture Notes in Computer Science*, 145–156. Berlin: Springer.

Friedman, J. H. and Stuetzle, W. (1981) Projection pursuit regression. *Journal of the American Statistical Association*, **76**, 817–823.

Fukumizu, K. (2000) Statistical active learning in multilayer perceptrons. *IEEE Transactions on Neural Networks*, **11**, 17–26.

Funahashi, K. (1989) On the approximate realization of continuous mappings by neural networks. *Neural Networks*, **2**, 183–192.

Funahashi, K. (1998) Multilayer neural networks and Bayes decision theory. *Neural Networks*, **11**, 209–213.

Fürnkranz, J. and Hüllermeier, E. (2003) Pairwise preference learning and ranking. In *Machine Learning: 14th European Conference on Machine Learning, Cavtat-Dubrovnik, Croatia, September 22-26, 2003, Proceedings*, vol. 2837 of *Lecture Notes in Computer Science*, 145–156. Berlin: Springer.

Gautam, S., Kimeldorf, G. and Sampson, A. R. (1996) Optimized scorings for ordinal data for the general linear model. *Statistics & Probability Letters*, **27**, 231–239.

Genter, F. C. and Farewell, V. T. (1985) Goodness-of-link testing in ordinal regression models. *Canadian Journal of Statistics*, **13**, 37–44.

Ghosh, M., Mukhopadhyay, N. and Sen, P. K. (1996) *Sequential Estimation*. Wiley Series in Probability and Statistics. New York: John Wiley & Sons.

Goodman, L. A. (1983) The analysis of dependence in cross-classifications having ordered categories, using log-linear models for frequencies and log-linear models for odds. *Biometrics*, **39**, 149–160.

Graham, R. L., Knuth, D. E. and Patashnik, O. (1994) *Concrete Mathematics: A Foundation for Computer Science (2nd Edition)*. Boston: Addison-Wesley.

Green, P. J. and Silverman, B. W. (1994) *Nonparametric regression and generalized linear models*. No. 58 in Monographs on Statistics and Applied Probability. London: Chapman & Hall.

Greenland, S. (1985) An application of logistic models to the analysis of ordinal responses. *Biometrical Journal*, **27**, 189–197.

Greenland, S. (1994) Alternative models for ordinal logistic regression. *Statistics in Medicine*, **13**, 1665–1677.

Grouven, U., Bergel, F. and Schulz, A. (1996) Implementation of linear and quadratic discriminant analysis incorporating costs of misclassification. *Computer Methods and Programs in Biomedicine*, **49**, 55–60.

Hagan, M. T. and Menhaj, M. B. (1994) Training feedforward networks with the marquardt algorithm. *IEEE Transactions on Neural Networks*, **5**, 989–993.

Halgamuge, S. K. and Wang, L. P., eds. (2005) *Classification and clustering for knowledge discovery. Selected papers based on the presentation at the international conference on fuzzy systems and knowledge discovery (FSKD) 2002.*, vol. 4 of *Studies in Computational Intelligence*. Berlin: Springer.

Hand, D. J. (1996) Statistics and the theory of measurement. *Journal of the Royal Statistical Society: Series A (Statistics in Society)*, **159**, 445–492.

Hanisch, J. U. (2005) Rounded responses to income questions. *Allgemeines Statistisches Archiv*, **89**, 39–48.

Hartigan, J. A. (1975) *Clustering Algorithms*. New York: John Wiley & Sons.

Hassibi, B. and Stork, D. G. (1993) Second order derivatives for network pruning: Optimal brain surgeon. In *Advances in Neural Information Processing Systems 5*, 164–171. Morgan-Kaufmann.

Hassibi, B., Stork, D. G. and Wolff, G. (1994) Optimal brain surgeon: Extensions and performance comparison. In *Advances in Neural Information Processing Systems 6* (eds. J. D. Cowan, G. Tesauro and J. Alspector), 263–270. Morgan-Kaufmann.

Hastie, T., Tibshirani, R. and Friedman, J. (2001) *The Elements of Statistical Learning*. New York: Springer.

Hastie, T. J., Botha, J. L. and Schnitzler, C. M. (1989) Regression with an ordered categorical response. *Statistics in Medicine*, **8**, 785–794.

Hastie, T. J. and Tibshirani, R. (1987) Non-parametric logistic and proportional odds regression. *Applied Statistics*, **36**, 260–276.

Haykin, S. S. (1999) *Neural Networks: A Comprehensive Foundation*. Prentice Hall, 2 edn.

Hecht-Nielsen, R. (1987) Kolmogorov's mapping neural network existence theorem. In *International Conference on Neural Networks* (ed. M. Caudill), vol. 3, 11–13. IEEE Press.

Hecht-Nielsen, R. (1989) Theory of the backpropagation neural network. In *International Joint Conference on Neural Networks*, vol. 1, 593–605. IEEE Press.

Hecht-Nielsen, R. (1990) *Neurocomputing*. Reading: Addison-Wesley.

Heitjan, D. F. (1989) Inference from grouped continuous data: A review. *Statistical Science*, **4**, 164–183.

Heitjan, D. F. and Rubin, D. B. (1991) Ignorability and coarse data. *The Annals of Statistics*, **19**, 2244–2253.

Herbrich, R., Graepel, T. and Obermayer, K. (1999a) Regression models for ordinal data: A machine learning approach. Tech. Rep. TR 99–3, Technical University of Berlin.

Herbrich, R., Graepel, T. and Obermayer, K. (1999b) Support vector learning for ordinal regression. In *Artificial Neural Networks, Ninth International Conference on*, 97–102.

Heskes, T. and Wiegerinck, W. (1996) A theoretical comparison of batch-mode, on-line, cyclic, and almost-cyclic learning. *IEEE Transactions on Neural Networks*, **7**, 919–925.

Hornik, K. (1991) Approximation capabilities of multilayer feedforward networks. *Neural Networks*, **4**, 251–257.

Hornik, K., Stinchcombe, M. and White, H. (1989) Multilayer feedforward networks are universal approximators. *Neural Networks*, **2**, 359–366.

Hornik, K., Stinchcombe, M., White, H. and Auer, P. (1994) Degree of approximation results for feedforward networks approximating unknown mappings and their derivatives. *Neural Computation*, **6**, 1262–1275.

Hsiao, C. and Mountain, D. (1985) Estimating the short-run income elasticity of demand for electricity by using cross-sectional categorized data. *Journal of the American Statistical Association*, **80**, 259–265.

Huang, G.-B., Zhu, Q.-Y. and Siew, C.-K. (2006) Real-time learning capability of neural networks. *IEEE Transactions on Neural Networks*, **17**, 863–878.

Johnson, T. R. (2006) Generalized linear models with ordinally-observed covariates. *British Journal of Mathematical and Statistical Psychology*, **59**, 275–300. Kopie.

Jones, L. K. (1990) Constructive approximations for neural networks by sigmoidal functions. *Proceedings of the IEEE*, **78**, 1586–1589.

Jones, L. K. (1992) A simple lemma on greedy approximation in Hilbert space and convergence rates for projection pursuit regression and neural network training. *The Annals of Statistics*, **20**, 608–613.

Judge, G. G., Griffiths, W. E., Carter Hill, R., Lütkepohl, H. and Lee, T.-C. (1985) *The Theory and Practice of Econometrics*. Wiley Series in Probability and Mathematical Statistics. New York: John Wiley & Sons, 2 edn.

Kampen, J. and Swyngedouw, M. (2000) The ordinal controversy revisited. *Quality & Quantity*, **34**, 87–102.

Katsuura, H. and Sprecher, D. A. (1994) Computational aspects of Kolmogorov's superposition theorem. *Neural Networks*, **7**, 455–461.

Kauermann, G. (2000) Modeling longitudinal data with ordinal response by varying coefficients. *Biometrics*, **56**, 692–698.

Khurshid, A. and Sahai, H. (1993) Scales of measurements: An introduction and a selected bibliography. *Quality & Quantity*, **27**, 303–324.

Kianifard, F. and Swallow, W. H. (1996) A review of the development and application of recursive residuals in linear models. *Journal of the American Statistical Association*, **91**, 391–400.

Knapp, T. R. (1990) Treating ordinal scales as interval scales: An attempt to resolve the controversy. *Nursing Research*, **39**, 121–123.

Kotsiantis, S. B. and Pintelas, P. E. (2004) A cost sensitive technique for ordinal classification problems. In *Methods and Applications of Artificial Intelligence, Third Helenic Conference on AI* (eds. G. A. Vouros and T. Panayiotopoulos), vol. 3025 of *Lecture Notes in Computer Science*, 220–229. Springer.

Kramer, S. (1996) Structural regression trees. In *Proceedings of the Thirteenth National Conference on Artificial Intelligence (AAAI-96)*, 812–819. AAAI Press/MIT Press. Bestellt.

Kramer, S., Widmer, G., Pfahringer, B. and de Groeve, M. (2000) Prediction of ordinal classes using regression trees. In *Foundations of Intelligent Systems, 12th International Symposium, ISMIS 2000, Charlotte, NC, USA, October 11-14, 2000, Proceedings* (ed. Z. W. e. a. Ras), vol. 1932 of *Lecture Notes in Computer Science*, 426–434. New York: Springer.

Kramer, S., Widmer, G., Pfahringer, B. and de Groeve, M. (2001) Prediction of ordinal classes using regression trees. *Fundamenta Informaticae*, **47**, 1–13.

Krantz, D. H., Luce, R. D., Suppes, P. and Tversky, A. (1971) *Foundations of Measurement*, vol. 1. New York: Academic Press.

Kůrková, V. (1992) Kolmogorov's theorem and multilayer neural networks. *Neural Networks*, **5**, 501–506.

Kuan, C.-M. and White, H. (1994) Artificial neural networks: An econometric perspective. *Econometric Reviews*, **13**, 1–103.

Kukuk, M. (2002) Indirect estimation of (latent) linear models with ordinal regressors. a monte carlo study and some empirical illustrations. *Statistical Papers*, **43**, 379–399.

Läärä, E. and Matthews, J. N. S. (1985) The equivalence of two models for ordinal data. *Biometrika*, **72**, 206–207.

Lall, R., Campbell, M. J., Walters, S. J. and Morgan, K. (2002) A review of ordinal regression models applied on health-related quality of life assessments. *Statistical Methods in Medical Research*, **11**, 49–67.

Langley, P. (1995) Order effects in incremental learning. In *Learning in humans and machines: Towards an Interdisciplinary Learning Science. Pergamon* (ed. P. R. . H. Spada). Elsevier.

Le Cun, Y., Denker, J. S. and Solla, S. A. (1990) Optimal brain damage. In *Advances in Neural Information Processing Systems: Proceedings of the 1989 Conference* (ed. D. S. Touretzky), 598–605. San Mateo: Morgan-Kaufmann.

Lee, M. (1992) Median regression for ordered discrete response. *Journal of Econometrics*, **51**, 59–77.

Lee, T., White, H. and Granger, C. (1993) Testing for neglected nonlinearity in time series models. *Journal of Econometrics*, **56**, 269–290.

Lehmann, E. L. (1983) *Theory of Point Estimation*. Wiles Series in Probability and Mathematical Statistics. New York: John Wiley & Sons.

Lehmann, E. L. (1986) *Testing Statistical Hypotheses*. Wiles Series in Probability and Mathematical Statistics. New York: John Wiley & Sons, 2 edn.

Leshno, M., Lin, V. Y., Pinkus, A. and Schocken, S. (1993) Multilayer feedforward networks with a nonpolynomial activation function can approximate any function. *Neural Networks*, **6**, 861–867.

Lewicki, G. and Marino, G. (2004) Approximation of functions of finite variation by superpositions of a sigmoidal function. *Applied Mathematics Letters*, **17**, 1147–1152.

Li, L. and Lin, H.-T. (2006) Ordinal regression by extended binary classification. In *Advances in neural information processing systems (NIPS)* (eds. B. Schölkopf, J. Platt and T. Hoffman), 865–872. Cambridge, MA: MIT Press. URL http://books.nips.cc/papers/files/nips19/NIPS2006_0880.pdf.

Lipsitz, S. R., Fitzmaurice, G. M. and Molenberghs, G. (1996) Goodness-of-fit tests for ordinal response regression models. *Applied Statistics*, **45**, 175–190.

Liu, I. and Agresti, A. (2005) The analysis of ordered categorical data: An overview and a survey of recent developments. *Sociedad de Estadística e Investigatión Operativa Test*, **14**, 1–73.

Liu, L. C. and Hedeker, D. (2006) A mixed-effects regression model for longitudinal multivariate ordinal data. *Biometrics*, **62**, 261–268.

Lord, F. M. (1953) On the statistical treatment of football numbers. *American Psychologist*, **8**, 750–751.

Luce, R. D. (1959) On the possible psychophysical laws. *Psychological Review*, **66**, 81–95.

Luce, R. D. (1990) "on the possible psychophysical laws" revisited: Remarks on cross-modal matching. *Psychological Review*, **97**, 66–77.

MacDonald, K. I. (1973) "ordinal regression?" a comment. *American Sociological Review*, **38**, 494–495.

Marcus-Roberts, H. M. and Roberts, F. S. (1987) Meaningless statistics. *Journal of Educational Statistics*, **12**, 338–394.

Marquardt, D. W. (1963) An algorithm for least-squares estimation of nonlinear parameters. *Journal of the Society for Industrial and Applied Mathematics*, **11**, 431–441.

Marshall, R. J. (1999) Classification to ordinal categories using a search partition methodology with an application in diabetes screening. *Statistics in Medicine*, **18**, 2723–2735.

Mathieson, M. J. (1996) Ordinal models for neural networks. In *Neural Networks in Financial Engineering. Proceedings of the Third International Conference on Neural Networks in the Capital Markets* (eds. A.-P. N. Refenes, Y. Abu-Mostafa, J. Moody and A. Weigend), 523–536. Singapore: World Scientific.

Mayer, L. S. (1971) A note on treating ordinal data as interval data. *American Sociological Review*, **36**, 519–520.

McCullagh, P. (1980) Regression models for ordinal data. *Journal of the Royal Statistical Society: Series B (Methodological)*, **42**, 109–142.

McCullagh, P. and Nelder, J. H. (1989) *Generalized Linear Models*. London: Chapman & Hall, 2 edn.

McHugh, R. B. (1963) Comment on "scales and statistics: Parametric and nonparametric". *Psychological Bulletin*, **60**, 350.

Mendil, B. and Benmahammed, K. (1999) Simple activation functions for neural and fuzzy neural networks. In *Proceedings of the 1999 IEEE International Symposium on Circuits and Systems*, vol. 5. IEEE Press.

Mhaskar, H. N. (1993) Approximation properties of a multi-layer feedforward artificial neural network. *Advances in Computational Mathematics*, **1**, 61–80.

Mhaskar, H. N. and Micchelli, C. A. (1992) Approximation by superposition of sigmoidal and radial basis functions. *Advances in Applied Mathematics*, **13**, 350–373.

Mhaskar, H. N. and Micchelli, C. A. (1994) How to choose an activation function. In *Advances in Neural Information Processing Systems* (eds. J. D. Cowan, G. Tesauro and J. Alspector), 319–326. Morgan-Kaufmann.

Michell, J. (1986) Measurement scales and statistics: A clash of paradigms. *Psychological Bulletin*, **100**, 398–407.

Molenberghs, G. and Verbeke, G. (2004) Meaningful statistical model formulations for repeated measures. *Statistica Sinica*, **14**, 989–1020.

Morgan, B. J. T. (1984) *Elements of Simulation*. Chapman & Hall.

Murata, N. (1998) A statistical study of on-line learning. In *On-line learning in neural networks* (ed. D. Saad), chap. 4, 63–92. New York: Cambridge University Press.

Murata, N., Kawanabe, M., Ziehe, A., Müller, K.-R. and ichi Amari, S. (2002) On-line learning in changing environments with applications in supervised and unsupervised learning. *Neural Networks*, **15**, 743–760.

Murata, N., Müller, K. R., Ziehe, A. and Amari, S. (1997) Adaptive on-line learning in changing environments. In *Advances in Neural Information Processing Systems* (eds. M. C. Mozer, M. I. Jordan and T. Petsche), vol. 9, 599–605.

Nakamura, M., Mines, R. and Kreinovich, V. (1993) Guaranteed intervals for Kolmogorov's theorem (and their possible relation to neural networks). *Interval Computations*, **3**, 183–199.

Narens, L. (1985) *Abstract measurement theory*. MIT Press Cambridge, Mass.

Nelder, J. A. (1990) The knowledge needed to computerise the analysis and interpretation of statistical information. In *Expert Systems and Artificial Intelligence: The Need for Information about Data* (ed. M. C. Fessey), 23–27. London: Library Association.

Nelder, J. A. and Wedderburn, R. W. M. (1972) Generalized linear models. *Journal of the Royal Statistical Society: Series A (General)*, **135**, 370–384.

Norris, C. M., Ghali, W. A., Saunders, L. D., Brant, R., Galbraith, D., Faris, P. and Knudtson, M. L. (2006) Ordinal regression model and the linear regression model were superior to the logistic regression models. *Journal of Clinical Epidemiology*, **59**, 448–456.

O'Connell, A. A. (2006) *Logistic Regression Models for Ordinal Response Variables*. No. 146 in Quantitative applications in the social sciences. Thousand Oaks, Calif.

Peterson, B. and Harrell, Jr, F. E. (1990) Partial proportional odds models for ordinal response variables. *Applied Statistics*, **39**, 205–217.

Piccarreta, R. (2004) Ordinal classification trees based on impurity measures. In *Advances in Multivariate Data Analysis: Proceedings Of The Meeting Of The Classification And Data Analysis Group (CLADAG) of the Italian Statistical Society* (eds. H.-H. Bock, M. Chiodi and A. Mineo), Studies in Classification, Data Analysis, and Knowledge Organization, 39–51. Berlin: Springer.

Potharst, R. and Bioch, J. C. (1999) A decision tree algorithm for ordinal classification. In *Advances in Intelligent Data Analysis: Third International Symposium, IDA-99, Amsterdam, the Netherlands, August 9-11, 1999, Proceedings* (eds. D. J. Hand, J. N. Kok and M. R. Berthold), vol. 1642 of *Lecture Notes in Computer Science*, 187–198. Berlin: Springer.

Potharst, R. and Bioch, J. C. (2000) Decision trees for ordinal classification. *Intelligent Data Analysis*, **4**, 97–111.

Potharst, R. and van Wezel, M. (2005) Generating artificial data with monotonicity constraints. Tech. Rep. EI 2005-06, Erasmus University Rotterdem. URL http://hdl.handle.net/1765/1916.

Press, S. J. and Wilson, S. (1978) Choosing between logistic regression and discriminant analysis. *Journal of the American Statistical Association*, **73**, 699–705.

Pulkstenis, E. and Robinson, T. J. (2004) Goodness-of-fit tests for ordinal response regression models. *Statistics in Medicine*, **23**, 999–1014.

Quinn, M. (2003) *Parallel Programming in C with MPI and OpenMP*. Boston: McGraw-Hill.

Rajaram, S., Garg, A., Zhou, X. S. and Huang, T. S. (2003) Classification approach towards banking and sorting problems. In *Machine Learning: 14th European Conference on Machine Learning, Cavtat-Dubrovnik, Croatia, September 22-26, 2003, Proceedings* (eds. N. Lavrac, D. Gamberger, L. Todorovski and H. Blockeel), vol. 2837 of *Lecture Notes in Computer Science*, 301–312. Berlin: Springer.

Rao, C. R. (1973) *Linear Statistical Inference and its Applications*. Wiles Series in Probability and Mathematical Statistics. New York: John Willey & Sons, 2 edn.

Reed, R. D. and Marks, R. J. (1998) *Neural Smithing: Supervised Learning in Feedforward Artificial Neural Networks*. MIT Press Cambridge, MA, USA.

Refenes, A.-P. N. and Holt, W. T. (2001) Forecasting volatility with neural regression: A contribution to model adequacy. *IEEE Transactions on Neural Networks*, **12**, 850–864.

Richard, M. D. and Lippmann, R. P. (1991) Neural network classifiers estimate bayesian a posteriori probabilities. *Neural Computation*, **3**, 461–483.

Ripley, B. D. (1993) Statistical aspects of neural networks. In *Networks and Chaos—Statistical and Probabilistic Aspects* (eds. O. E. Barndorff-Nielsen, J. L. Jensen and W. S. Kendall), vol. 50 of *Monographs on Statistics and Applied Probability*, chap. 2, 40–123. London: Chapman & Hall.

Ripley, B. D. (1994a) Network methods in statistics. In *Probability, Statistics and Optimisation* (eds. F. P. Kelly and P. Whittle), Wiley Series in Probability and Mathematical Statistics, chap. 19, 241–253. Chichester: Wiley.

Ripley, B. D. (1994b) Neural networks and related methods for classification. *Journal of the Royal Statistical Society: Series B (Methodological)*, **56**, 409–456.

Ripley, B. D. (1995) Statistical ideas for selecting network architectures. In *Neural Networks: Artificial Intelligence and Industrial Applications* (eds. B. Kappen and S. Gielen), 183–190. London: Springer.

Ripley, B. D. (1996) *Pattern recognition and neural networks*. Cambridge Univ. Press.

Ripley, B. D. (1997) Can statistical theory help us use neural networks better? In *Interface 97. 29th Symposium on the Interface: Computing Science and Statistics*.

Ripley, B. D. and Mardia, K. V. (1994) Neural networks and flexible regression and discrimination. *Journal of Applied Statistics*, **21**, 39–58.

Robbins, H. and Monro, S. (1951) A stochastic approximation method. *The Annals of Mathematical Statistics*, **22**, 400–407.

Roberts, F. S. (1979) *Measurement theory*. Addison-Wesley.

Rojas, R. (1996) *Theorie der neuronalen Netze*. Berlin: Springer.

Ronning, G. and Kukuk, M. (1996) Efficient estimation of ordered probit models. *Journal of the American Statistical Association*, **91**, 1120–1129.

Rosenblatt, F. (1958) The perceptron: A probabilistic model for information storage and organization in the brain. *Psychological Review*, **65**, 386–408.

Rosenblatt, F. (1962) *Principles of Neurodynamics: Perceptrons and the Theory of Brain Mechanisms*. Spartan Books.

Ruck, D. W., Rogers, S. K., Kabrisky, M., Oxley, M. E. and Suter, B. W. (1990) The multilayer perceptron as an approximation to a Bayes optimal discriminant function. *IEEE Transactions on Neural Networks*, **1**, 296–298.

Rudolfer, S. M., Watson, P. C. and Lesaffre, E. (1995) Are ordinal models useful for classification? a revised analysis. *Journal of Statistical Computation and Simulation*, **52**, 105–132.

Rumelhart, D. E., Hinton, G. E. and Williams, R. J. (1986a) Learning internal representations by error propagation. In (Rumelhart and McClelland, 1986), chap. 8, 318–362.

Rumelhart, D. E., Hinton, G. E. and Williams, R. J. (1986b) Learning representations by back-propagating errors. *Nature*, **323**, 533–536.

Rumelhart, D. E. and McClelland, J. L., eds. (1986) *Parallel Distributed Processing - Explorations in the Microstructure of Cognition*, vol. 1. Cambridge: MIT Press.

Saad, D. (1999) *On-Line Learning in Neural Networks*. Cambridge University Press.

Saad, D. and Rattray, M. (1998) Optimal on-line learning in multilayer neural networks. In *On-line learning in neural networks* (ed. D. Saad), chap. 7, 135–164. New York: Cambridge University Press.

Sarle, W. S. (1994) Neural networks and statistical models. In *Proceedings of the Nineteenth Annual SAS Users Group International Conference*, 1538–1550. Cary: SAS Institute.

Sarle, W. S. (1997a) Measurement theory: Frequently asked questions. URL ftp://ftp.sas.com/pub/neural/measurement.html.

Sarle, W. S. (1997b) Neural network FAQ, part 2 of 7: Sequential learning, catastrophic interference, and the stability-plasticity dilemma. URL ftp://ftp.sas.com/pub/neural/FAQ2.html#A_styles_sequential.

Scharfstein, D. O., Liang, K.-Y., Eaton, W. and Chen, L.-S. (2001) The quadratic cumulative odds regression model for scored ordinal outcomes: application to alcohol dependence. *Biostatistics*, **2**, 473–483.

Schiffers, J. (1997) A classification approach incorporating misclassification costs. *Intelligent Data Analysis*, **1**, 59–68.

Schumacher, M., Roßner, R. and Vach, W. (1996) Neural networks and logistic regression: Part I. *Computational Statistics & Data Analysis*, **21**, 661–682.

Seber, G. A. F. and Wild, C. J. (2003) *Nonlinear Regression*. Wiles Series in Probability and Statistics. Hoboken: John Wiley & Sons.

Sen, A. K. and Srivastava, M. (1990) *Regression Analysis: Theory, Methods, and Applications*. New York: Springer.

Shashua, A. and Levin, A. (2002) Taxonomy of large margin principle algorithms for ordinal regression problems. Tech. Rep. 39, Leibniz Center for Research, School of Computer Science and Eng., Jerusalem.

Shen, L. and Joshi, A. K. (2005a) Flexible margin selection for reranking with full pairwise samples. In *Natural Language Processing: Second International Joint Conference, Jeju Island, Korea, October 11-13, 2005, Proceedings* (eds. K.-Y. Su, J. ichi Tsujii, J.-H. Lee and O. Y. Kwong), vol. 3248 of *Lecture Notes in Computer Science*, 446–455. Berlin: Springer.

Shen, L. and Joshi, A. K. (2005b) Ranking and reranking with perceptron. *Machine Learning*, **60**, 73–96.

Si, J., Nelson, B. J. and Runger, C. G. (2003) Artificial neural network models for data mining, In *The Handbook of Data Mining* (ed. N. Ye), chap. 3, 41–66. Lawrence Erlbaum.

Simon, G. (1974) Alternative analyses for the singly-ordered contingency table. *Journal of the American Statistical Association*, **69**, 971–976.

Snell, E. J. (1964) A scaling procedure for ordered categorical data. *Biometrics*, **20**, 592–607.

Sprecher, D. A. (1965a) On the structure of continuous functions of several variables. *Transactions of the American Mathematical Society*, **115**, 340–355.

Sprecher, D. A. (1965b) A representation theorem for continuous functions of several variables. *Proceedings of the American Mathematical Society*, **16**, 200–203.

Sprecher, D. A. (1966) On the structure of representations of continuous functions of several variables as finite sums of continuous functions of one variable. *Proceedings of the American Mathematical Society*, **17**, 98–105.

Sprecher, D. A. (1993) A universal mapping for Kolmogorov's superposition theorem. *Neural Networks*, **6**, 1089–1094.

Sprecher, D. A. (1996a) A numerical construction of a universal function for Kolmogorov's superpositions. *Neural Network World*, **6**, 711–718.

Sprecher, D. A. (1996b) A numerical implementation of Kolmogorov's superpositions. *Neural Networks*, **9**, 765–772.

Stevens, S. S. (1946) On the theory of scales of measurement. *Science*, **103**, 677–680.

Stevens, S. S. (1951) Mathematics, measurement, and psychophysics. In *Handbook of Experimental Psychology* (ed. S. S. Stevens). John Wiley, 8 edn. Psych. ST 24/66.

Stevens, S. S. (1955) On the averaging of data. *Science*, **121**, 113–116.

Stevens, S. S. (1959) Measurement, psychophysics, and utility. In *Measurement: Definitions and Theories* (eds. C. W. Churchman and P. Ratoosh), chap. 2, 18–63. John Wiley.

Stevens, S. S. (1968) Measurement, statistics, and the schemapiric view. *Science*, **161**, 849–856.

Stewart, M. B. (2005) A comparison of semiparametric estimators for the ordered response model. *Computational Statistics & Data Analysis*, **49**, 555–573.

Stiger, T. R., Barnhart, H. X. and Williamson, J. M. (1999) Testing proportionality in the proportional odds model fitted with GEE. *Statistics in Medicine*, **18**, 1419–1433.

Stinchcombe, M. and White, H. (1989) Universal approximation using feedforward networks with non-sigmoid hidden layer activation functions. In *International Joint Conference on Neural Networks*, vol. 1, 613–617. IEEE Press.

Sundararajan, N. and Saratchandran, P. (1998) *Parallel Architectures for Artificial Neural Networks*. IEEE Computer Society Press.

Tan, P.-N., Steinbach, M. and Kumar, V. (2005) *Introduction to Data Mining*. Boston: Addison-Wesley.

Taylor, A. B., West, S. G. and Aiken, L. S. (2006) Loss of power in logistic, ordinal logistic, and probit regression when an outcome variable is coarsely categorized. *Educational and Psychological Measurement*, **66**, 228–239.

Taylor, G. W. and Becker, M. P. (1998) Increased efficiency of analyses: cumulative logistic regression vs. ordinary logistic regression. *Community Dentistry and Oral Epidemiology*, **26**, 1–6.

Teräsvirta, T., Lin, C. and Granger, C. (1993) Power of the neural network linearity test. *Journal of Time Series Analysis*, **14**, 209–220.

Terza, J. V. (1987) Estimating linear models with ordinal qualitative regressors. *Journal of Econometrics*, **34**, 275–291.

Thimm, G., Fiesler, E. and Moerland, P. (1995) Gain elimination from backpropagation neural networks. In *International Conference on Neural Networks*, vol. 1, 365–368. IEEE Press. Besorgen schwierig.

Thimm, G., Moerland, P. and Fiesler, E. (1996) The interchangeability of learning rate and gain in backpropagation neural networks. *Neural Computation*, **8**, 451–460.

Thompson, R. and Baker, R. J. (1981) Composite link functions in generalized linear models. *Applied Statistics*, **30**, 125–131.

Torra, V., Domingo-Ferrer, J., Mateo-Sanz, J. M. and Ng, M. (2006) Regression for ordinal variables without underlying continuous variables. *Information Sciences*, **176**, 465–474.

Tørresen, J. and Tomita, S. (1998) A review of parallel implementations of backpropagation neural networks, In *Parallel Architectures for Artificial Neural Networks: Paradigms and Implementations* (eds. N. Sundararajan and P. Saratchandran), chap. 2, 25–63. IEEE CS Press.

Townsend, J. T. and Ashby, F. G. (1984) Measurement scales and statistics: The misconception misconceived. *Psychological Bulletin*, **96**, 394–401.

Tutz, G. (1990) Sequential item response models with an ordered response. *British Journal of Mathematical and Statistical Psychology*, **43**, 39–55.

Tutz, G. (1991) Sequential models in categorical regression. *Computational Statistics & Data Analysis*, **11**, 275–295.

Tutz, G. (2003) Generalized semiparametrically structured ordinal models. *Biometrics*, **59**, 263–273.

Tutz, G. (2005) Modelling of repeated ordered measurements by isotonic sequential regression. *Statistical Modelling*, **5**, 269–287.

Tutz, G. and Hennevogl, W. (1996) Random effects in ordinal regression models. *Computational Statistics & Data Analysis*, **22**, 537–557.

Uebersax, J. S. (1999) Probit latent class analysis with dichotomous or ordered category measures: Conditional independence/dependence models. *Applied Psychological Measurement*, **23**, 283.

Vach, W., Roßner, R. and Schumacher, M. (1996) Neural networks and logistic regression: Part II. *Computational Statistics & Data Analysis*, **21**, 683–701.

Vapnik, V. N. (1998) *Statistical Learning Theory.* New York: Wiley-Interscience.

Velleman, P. F. and Wilkinson, L. (1993) Nominal, ordinal, interval, and ratio typologies are misleading. *The American Statistician*, **47**, 65–72.

Walker, S. H. and Duncan, D. B. (1967) Estimation of the probability of an event as a function of several independent variables. *Biometrika*, **54**, 167–179.

Walter, S. D., Feinstein, A. R. and Wells, C. K. (1987) Coding ordinal independent variables in multiple regression analyses. *American Journal of Epidemiology*, **125**, 319–323.

Wang, S. (1995) The unpredictability of standard back propagation neural networks in classification applications. *Management Science*, **41**, 555–559.

Wang, Y. J. (1986) Order-dependent parameterization of multinomial distributions. *Scandinavian Journal of Statistics*, **13**, 199–205.

Warner, B. and Misra, M. (1996) Understanding neural networks as statistical tools. *The American Statistician*, **50**, 284–293.

White, H. (1989a) An additional hidden unit test for neglected nonlinearity in multilayer feedforward networks. *Neural Networks, International Joint Conference on*, 451–455.

White, H. (1989b) Learning in artificial neural networks: A statistical perspective. *Neural Computation*, **1**, 425–464.

White, H. (1989c) Some asymptotic results for learning in single hidden-layer feedforward network models. *Journal of the American Statistical Association*, **84**, 1003–1013.

White, H. (1994) Parametric statistical estimation with artificial neural networks: A condensed discussion. In *From Statistics to Neural Networks* (eds. V. Cherkassky, J. H. Friedman and H. Wechsler), vol. 136 of *Computer and Systems Sciences*, 127–146. Berlin: Springer.

Widrow, B. and Hoff, M. E. (1960) Adaptive switching circuits. *IRE WESCON Convention Record*, 96–104.

Williams, O. D. and Grizzle, J. E. (1972) Analysis of contingency tables having ordered response categories. *Journal of the American Statistical Association*, **67**, 55–63.

Woodward, T. S., Hunter, M. A. and Kadlec, H. (2002) The comparative sensitivity of ordinal multiple regression and least squares regression to departures from interval scaling. *British Journal of Mathematical and Statistical Psychology*, **55**, 305–315.

Zell, A. (1994) *Simulation neuronaler Netze*. München: Addison Wesley.

Zhang, G. P. (2000) Neural networks for classification: A survey. *IEEE Transactions on Systems, Man and Cybernetics, Part C: Applications and Reviews*, **30**, 451–462.

Zhu, J. and Sutton, P. (2003) FPGA implementations of neural networks – A survey of a decade of progress. In *FPL* (eds. P. Y. K. Cheung, G. A. Constantinides and J. T. de Sousa), vol. 2778 of *Lecture Notes in Computer Science*, 1062–1066. Springer.

A Additional Theorems

A.1 Theorems Concerning Equivalence Classes for Measurements of Scale

The following theorems are intended to prove the existence and the completeness of the equivalence classes related to different measurements of scale described in subsection 2.1.1.

Theorem A.1

The set of injective mappings is the equivalence class related to the nominal scale.

Proof. For every mapping we have $\forall_{a,b \in \mathbb{R}} a = b \rightarrow f(a) = f(b)$. It is easy to see that $a, b \in M a \sim b \rightarrow f(\psi(a)) = f(\psi(b))$. Thus, every mapping preserves the properties i, ii and iii.

Restricting the set of mappings to injective mappings ensures $f(\psi(a)) = f(\psi(b)) \rightarrow \psi(a) = \psi(b)$. $\qquad\square$

Theorem A.2

The set of strictly isotone functions is the equivalence class related to the ordinal scale.

Proof. Let $a, b, c \in M$ and $a \succeq b$ and $b \succeq c$. Then we have $\psi(a) \geq \psi(b) \geq \psi(c)$. As f is strictly isotone we also have $f(\psi(a)) \geq f(\psi(b)) \geq f(\psi(c))$. Thus, strictly isotone functions preserve the properties i to v.

Let us assume a function f with $a, b \in M$ with $\psi(a) \gneq psi(b)$ and $f(\psi(a)) = f(\psi(b))$, i.e. a function that is not surjective. This would imply that $\psi(a) \leq \psi(b)$ and lead to a contradiction. If there were an f that is strictly monotone decreasing, we could find $a, b \in M$ with $\psi(a) \geq \psi(b)$ but $f(\psi(a)) < f(\psi(b))$. As strictly monotone increasing or decreasing functions are the only two possibilities for surjective functions on $\mathbb{R}$, the class of strictly isotone functions is complete. $\qquad\square$

Theorem A.3

The set of functions with $f(x; a, b) = g(ag^{-1}(x) + b)$ with $a, b \in \mathbb{R}$, $a > 0$ and g convex and strictly isotone is the equivalence class related for the g^{-1}-interval scale.

Proof. All functions of the form $f(x; a, b) = g(ag^{-1}(x)+b)$ with $a, b \in \mathbb{R}$, $a > 0$ are strictly isotone with g convex and strictly isotone. Therefore, the properties i to v are given with A.2. Furthermore, we get for $m_1, m_2, m_3, m_4 \in M$ with $g^{-1}(\psi(m_1)) - g^{-1}(\psi(m_2)) \geq g^{-1}(\psi(m_3)) - g^{-1}(\psi(m_4))$:

$$
\begin{aligned}
g^{-1}\Big(f(\psi(m_1))\Big) - g^{-1}\Big(f(\psi(m_2))\Big) &= a\left[g^{-1}(\psi(m_1)) - g^{-1}(\psi(m_2))\right] \\
&\geq a\left[g^{-1}(\psi(m_3)) - g^{-1}(\psi(m_4))\right] \\
&= g^{-1}\Big(f(\psi(m_3))\Big) - g^{-1}\Big(f(\psi(m_4))\Big)
\end{aligned}
$$

All functions in the equivalence class are of the general form $f = g \circ h \circ g^{-1}$. Then the function h has to be linear, if it was not, we could find $m_1 \succeq m_2 \succeq m_3 \in M$ for a strict convex and isotone h with

$$
\begin{aligned}
&g^{-1}(\psi(m_3)) - g^{-1}(\psi(m_2)) = g^{-1}(\psi(m_2)) - g^{-1}(\psi(m_1)) \\
\Leftrightarrow\ &\frac{1}{2}g^{-1}(\psi(m_3)) + g^{-1}(\psi(m_1)) = g^{-1}(\psi(m_2)) \\
\Leftrightarrow\ &g^{-1}\Big(h\big(\tfrac{1}{2}\psi(m_3) + \psi(m_1)\big)\Big) > g^{-1}\Big(h(\psi(m_2))\Big) \\
\Leftrightarrow\ &\frac{1}{2}\psi(m_3) + \psi(m_1) > \psi(m_2)
\end{aligned}
$$

producing a contradiction. h has to be isotone because f has to be isotone. For concave h the proof is analogue. $\qquad\square$

Theorem A.4

The set of linear functions without a constant is the equivalence class related to the ratio scale.

Proof. As the ratio scale is a combination of the interval and the log-interval scale for which we have proven the equivalence classes $ax + b$ resp. ax^b the equivalence class related to the ratio scale has to be the intersection of these two, thus the class ax with $a \in \mathbb{R}$ constant. $\qquad\qquad\qquad\qquad\square$

A.2 Calculations with Series

Theorem A.5

*Let $(a_i) = \sum_{n=0}^{\infty} a_{i,n}$ be a family of absolute convergent series with limits A_i.[98]
Then the Cauchy product can be generalized to all $m \in \mathbb{N}$*

$$\prod_{i=0}^{m} \left(\sum_{n=0}^{\infty} a_{i,n} \right) = \sum_{n=0}^{\infty} \sum_{\substack{\sum_{j=0}^{m} k_j = n \\ k_j \geq 0}} \left(\prod_{j=0}^{m} a_{j,k_j} \right) \qquad (A.1)$$

is absolute convergent and its limit is $\prod_{i=0}^{m} A_i$.

Proof. The convergence of the product can trivially be proved by Merten's theorem.

The formula can be proved by using mathematical induction over m:

1. With the Cauchy product of series it is sure:

$$\sum_{n=0}^{\infty} a_{0,n} \cdot \sum_{n=0}^{\infty} a_{1,n} = \sum_{n=0}^{\infty} \sum_{k=0}^{n} a_{0,k} a_{1,n-k} = \sum_{n=0}^{\infty} \sum_{\substack{\sum_{j=0}^{1} k_j = n \\ k_j \geq 0}} \left(\prod_{j=0}^{1} a_{j,k_j} \right)$$

Thus, the formula (A.1) is correct for $m = 1$.

2. Let (A.1) be correct for a fixed $m \in \mathbb{N}$. Then

$$\prod_{i=0}^{m+1} \left(\sum_{n=0}^{\infty} a_{i,n} \right) = \left[\sum_{n=0}^{\infty} \sum_{\substack{\sum_{j=0}^{m} k_j = n \\ k_j \geq 0}} \left(\prod_{j=0}^{m} a_{j,k_j} \right) \right] \cdot \sum_{n=0}^{\infty} a_{m+1,n}$$

---

[98]It is sufficient that all series (a_i) except one of them are absolute convergent (Merten's Theorem).

and with

$$b_\ell := \sum_{\substack{\sum_{j=0}^{m} k_j = \ell \\ k_j \geq 0}} \left(\prod_{j=0}^{m} a_{j,k_j} \right)$$

follows with help of the simple Cauchy product:

$$\prod_{i=0}^{m+1} \left(\sum_{n=0}^{\infty} a_{i,n} \right) = \left[\sum_{n=0}^{\infty} b_n \right] \cdot \sum_{n=0}^{\infty} a_{m+1,n}$$

$$= \sum_{n=0}^{\infty} \sum_{\ell=0}^{n} b_\ell a_{m+1,n-\ell}$$

$$= \sum_{n=0}^{\infty} \sum_{\ell=0}^{n} \sum_{\substack{\sum_{j=0}^{m} k_j = \ell \\ k_j \geq 0}} \left(\prod_{j=0}^{m} a_{j,k_j} \right) a_{m+1,n-\ell}$$

it follows with $k_{m+1} = n - \ell$:

$$\prod_{i=0}^{m+1} \left(\sum_{n=0}^{\infty} a_{i,n} \right) = \sum_{n=0}^{\infty} \sum_{\ell=0}^{n} \sum_{\substack{\sum_{j=0}^{m} k_j = \ell \\ k_j \geq 0}} \left(\prod_{j=0}^{m+1} a_{j,k_j} \right)$$

$$= \sum_{n=0}^{\infty} \sum_{\substack{\sum_{j=0}^{m+1} k_j = n \\ k_j \geq 0}} \left(\prod_{j=0}^{m+1} a_{j,k_j} \right)$$

$\square$

Corollary A.6

Let $g(x) = \sum_{n=0}^{\infty} a_n x^n$ be an absolute convergent power series and $m \in \mathbb{N}$. Then

$$(g(x))^m = \left(\sum_{n=0}^{\infty} a_n x^n \right)^m = \sum_{n=0}^{\infty} \left(\sum_{\substack{\sum_{j=0}^{m} k_j = n \\ k_j \geq 0}} \prod_{j=0}^{m} a_{k_j} \right) x^n \qquad (A.2)$$

is a convergent power series with limit $(g(x))^m$.

Proof. Theorem A.5 and some algebra.   □

Corollary A.7 (composite power series)
Let $g(x) = \sum_{n=0}^{\infty} a_n x^n$ and $h(x) = \sum_{m=0}^{\infty} b_m x^m$ be two absolute convergent power series. Then the power series of the composite function

$$g \circ h(x) = \sum_{m=0}^{\infty} \left(\sum_{n=0}^{\infty} a_n \sum_{\substack{\sum_{j=0}^{m} k_j = n \\ k_j \geq 0}} \prod_{j=0}^{m} b_{k_j} \right) x^m$$

is absolute convergent.

Proof.

$$g \circ h(x) = \sum_{n=0}^{\infty} a_n \left(\sum_{m=0}^{\infty} b_m x^m \right)^n$$

using corollary A.6

$$= \sum_{n=0}^{\infty} a_n \left(\sum_{m=0}^{\infty} \left(\sum_{\substack{\sum_{j=0}^{m} k_j = n \\ k_j \geq 0}} \prod_{j=0}^{m} b_{k_j} \right) x^m \right)$$

Absolute convergence allows reordering to obtain the result.   □

Lemma A.8 (addition of two series)
Let $g(x|w) = \sum_{n=0}^{\infty} a_n \langle w, x \rangle^n$ and $h(x|z) = \sum_{m=0}^{\infty} b_m \langle z, x \rangle^m$ be two absolute convergent power series with d dimensional vectors x, w and z. Further let $c_1, c_2 \in \mathbb{R}$. Then the weighted sum of two power series $c_1 \cdot g(x) + c_2 \cdot h(x)$ is absolute convergent and given by

$$\sum_{n=0}^{\infty} \sum_{k_1, \ldots, k_d} \binom{n}{k_1, \ldots, k_d} \left(c_1 \cdot a_n \prod_{i=1}^{d} w_i^{k_i} + c_2 \cdot b_n \prod_{i=1}^{d} z_i^{k_i} \right) \prod_{i=1}^{d} x_i^{k_i}$$

Proof. Starting with

$$c_1 g(x) + c_2 h(x) = \sum_{n=0}^{\infty} c_1 a_n \langle \boldsymbol{w}, \boldsymbol{x} \rangle^n + \sum_{n=0}^{\infty} c_2 b_n \langle \boldsymbol{z}, \boldsymbol{x} \rangle^n$$

using absolute convergence for reordering the series,

$$= \sum_{n=0}^{\infty} c_1 a_n \langle \boldsymbol{w}, \boldsymbol{x} \rangle^n + c_2 b_n \langle \boldsymbol{z}, \boldsymbol{x} \rangle^n$$

using the multinomial theorem,

$$= \sum_{n=0}^{\infty} c_1 a_n \sum_{k_1,\ldots,k_d} \binom{n}{k_1,\ldots,k_d} \prod_{i=1}^{d} (w_i x_i)^{k_i}$$

$$+ c_2 b_n \sum_{k_1,\ldots,k_d} \binom{n}{k_1,\ldots,k_d} \prod_{i=1}^{d} (z_i x_i)^{k_i}$$

$$= \sum_{n=0}^{\infty} \sum_{k_1,\ldots,k_d} \left(c_1 a_n \binom{n}{k_1,\ldots,k_d} \prod_{i=1}^{d} w_i^{k_i} \right.$$

$$\left. + c_2 b_n \binom{n}{k_1,\ldots,k_d} \prod_{i=1}^{d} z_i^{k_i} \right) \prod_{i=1}^{d} x_i^{k_i}$$

$\square$

Theorem A.9 (weighted sum of series)
Let

$$\left(g_j(\boldsymbol{x}, \boldsymbol{w}_j) = \sum_{n=0}^{\infty} \tau_{j,n} \langle \boldsymbol{w}_j, \boldsymbol{x} \rangle^n \right)_{j \in \mathbb{N}}$$

be a family of absolute convergent power series in $\boldsymbol{x} \in \mathbb{R}^d$. *Let further* $(c_j)_{j \in \mathbb{N}}$

be a corresponding family of scalars. Then the weighted sum of the power series is absolute convergent and the resulting power series has the form:

$$\sum_j g_j(\boldsymbol{x}, \boldsymbol{w_j}) = \sum_{n=0}^{\infty} \sum_{k_1,...,k_d} \binom{n}{k_1, \ldots, k_d} \left(\sum_j c_j \tau_{j,n} \prod_{i=1}^{d} w_{ji}^{k_i} \right) \prod_{i=1}^{d} x_i^{k_i}$$

Proof. Using the result of lemma A.8 and simple induction over j.   □

B Lecture Evaluation Example

Answers are mostly given on an 5-point scale of agreement plus option for abstention; in the other cases, the scale is noted in brackets. In short, the following questions are included:

1 gender and program of study

    1.1 gender (options: male, female, n/a)

    1.2 current semester of study (options: 1-99)

    1.3 area of study (options: economics, business administration, information systems)

    1.4 degree, diploma or certificate (options: magister, diploma, bachelor, master)

2 course structure

    2.1 structure of the course content

    2.2 presentation of the course's learning goals

    2.3 instructor's organization of the course content

3 instructors's involvement

    3.1 motivation for the topic

    3.2 instructor's motivation in delivering the course content

    3.3 instructor's desire for knowledge transfer

4 delivery of course content

    4.1 instructor's oral expression capability

    4.2 quality of instructors's explanation of difficult ideas

    4.3 ease of following the lectures

5 self-assessment in relation to the course

    5.1 interest in course content

    5.2 motivation for self-study

    5.3 learning progress

6 media and materials

    6.1 quality of materials

    6.2 satisfaction with the use of media

7 learning conditions

    7.1 participant's behavior

    7.2 quality of the room

    7.3 proportion of missed classes
       (0-10%, 11-20%, 21-30%, 31-40%, >40%, n/s)

    7.4 proportion of canceled or stand-in classes (0-10%, 11-20%, 21-30%, 31-40%, >40%, n/s)

    7.5 academic level of the course

    7.6 grading of the course

8 proseminar evaluation

    8.1 attended proseminar (1-99)

    8.2 grading of the proseminar

9 self-assessment in relation to previous mathematical knowledge

    9.1 mathematics in school (options: basic course, basic course with marks, intensive course)

    9.2 grades in mathematics

209

EvaSys	Vorlesungsevaluation Wirtschaftswissenschaftliche Fakultät	vivid forms

6. Medien/Material

6.1 Wie zufrieden sind Sie mit den von der/dem Lehrenden zur Verfügung gestellten Materialien (Literaturhinweise, Reader, Onlinematerialien, Folien etc.)?   sehr zufrieden ☐ ☐ ☐ ☐ ☐ sehr unzufrieden   ☐ Enthaltung

6.2 Wie zufrieden sind Sie mit dem Einsatz von Medien (Beamer, Tafel etc.)?   sehr zufrieden ☐ ☐ ☐ ☐ ☐ sehr unzufrieden   ☐ Enthaltung

7. Rahmenbedingungen/Gesamtbewertung

7.1 Wie sehr wurden Sie durch Unruhe unter der Hörerschaft gestört?   überhaupt nicht ☐ ☐ ☐ ☐ ☐ sehr stark   ☐ Enthaltung

7.2 Wie geeignet fanden Sie den Raum für die Veranstaltung?   sehr gut ☐ ☐ ☐ ☐ ☐ sehr schlecht   ☐ Enthaltung

7.3 Wie häufig haben Sie gefehlt?   ☐ 0-10%   ☐ 11-20%   ☐ 21-30%
☐ 31-40%   ☐ >40%   ☐ Enthaltung

7.4 Wie häufig ist die Veranstaltung ausgefallen oder wurde nicht von der/dem Lehrenden selbst gehalten?   ☐ 0-10%   ☐ 11-20%   ☐ 21-30%
☐ 31-40%   ☐ >40%   ☐ Enthaltung

7.5 Das Niveau der Veranstaltung ist...   viel zu hoch ☐ ☐ ☐ ☐ ☐ viel zu niedrig   ☐ Enthaltung

7.6 Wenn man alles in einer Schulnote zusammenfasst, würde ich folgende Note geben: sehr gut ☐ ☐ ☐ ☐ ☐ mangelhaft   ☐ Enthaltung

7.7 Welche Verbesserungsvorschläge haben Sie? Was finden Sie besonders gut/schlecht?
(Hinweis: Schreiben Sie nur innerhalb des umrandeten Feldes. Die/der Lehrende erhält Ihren handschriftlichen Kommentar als Bildausschnitt.)

8. Bewertung des Proseminars

8.1 Ich besuche das Proseminar mit der Nummer (Bei häufigem Wechsel oder Enthaltung bitte 00 angeben)
10er ☐ ☐ ☐ ☐ ☐ ☐ ☐ ☐ ☐ ☐
1er ☐ ☐ ☐ ☐ ☐ ☐ ☐ ☐ ☐ ☐
x0 x1 x2 x3 x4 x5 x6 x7 x8 x9

8.2 Dem Proseminar gebe ich die Note   sehr gut ☐ ☐ ☐ ☐ ☐ mangelhaft   ☐ Enthaltung

9. Einschätzung der eigenen Mathematikvorkenntnisse

9.1 Mathematik in der gymnasialen Oberstufe hatte ich als
☐ Grundkurs   ☐ Grundkurs und Abiturfach   ☐ Leistungskurs

9.2 Meine Schulleistungen in Mathematik waren   sehr gut ☐ ☐ ☐ ☐ ☐ mangelhaft   ☐ Enthaltung

Index

Bisher erschienene und geplante Bände der Reihe
Advances in Information Systems and Management Science

ISSN 1611-3101

Bd. 1: Lars H. Ehlers

Content Management Anwendungen.

Spezifikation von Internet-Anwendungen
auf Basis von Content Management Systemen

ISBN 978-3-8325-0145-7   40.50 €

285 Seiten, 2003

Bd. 2: Stefan Neumann

Workflow-Anwendungen
in technischen Dienstleistungen.

Eine Referenz-Architektur für die Koordination
von Prozessen im Gebäude- und Anlagenmanagement

ISBN 978-3-8325-0156-3   40.50 €

310 Seiten, 2003

Bd. 3: Christian Probst

Referenzmodell für IT-Service-Informationssysteme.

ISBN 3-8325-0161-4   40.50 €

315 Seiten, 2003

Bd. 4: Jan vom Brocke

Referenzmodellierung.

Gestaltung und Verteilung von Konstruktionsprozessen

ISBN 978-3-8325-0161-7   40.50 €

424 Seiten, 2003

Bd. 5: Holger Hansmann

Architekturen Workflow-gestützter PPS-Systeme.

Referenzmodelle für die Koordination von Prozessen
der Auftragsabwicklung von Einzel- und Kleinserienfertigern

ISBN 978-3-8325-0282-9   40.50 €

299 Seiten, 2003

Bd. 17: Dietmar Stosik

Mehrzielorientierte Ablaufplanung bei auftragsorientierter Werkstattfertigung.

ISBN978-3-8325-0926-2   40.50 €
280 Seiten, 2003

Bd. 18: Rainer Babiel

Content Management in der öffentlichen Verwaltung.
Ein systemgestaltender Ansatz für die Justizverwaltung NRW

ISBN 978-3-8325-0927-9   40.50 €
226 Seiten, 2005

Bd. 19: Michael Ribbert

Gestaltung eines IT-gestützten Kennzahlensystems für das Produktivitätscontrolling operativer Handelsprozesse.
Ein fachkonzeptioneller Ansatz am Beispiel des klassischen Lagergeschäfts des Lebensmittelgroßhandels

ISBN 978-3-8325-0944-6   40.50 €
292 Seiten, 2005

Bd. 20: Peter Westerkamp

Flexible Elearning Platforms: A Service-Oriented Approach

ISBN 978-3-8325-1117-3   40.50 €
320 Seiten, 2005

Bd. 21: Timo Grabka

Referenzmodelle für Planspielplattformen.
Ein fachkonzeptioneller Ansatz zur Senkung der Konstruktions- und Nutzungskosten computergestützter Planspiele

ISBN 978-3-8325-1116-6   40.50 €
249 Seiten, 2006

Bd. 27: Volker Manthey

Controlling Integrierter Kampagnen - ein systemgestaltender Ansatz.

ISBN 978-3-8325-1373-3   40.50 €
244 Seiten, 2006

Bd. 28: Axel Winkelmann

Integrated Couponing. A Process-Based Framework for In-Store Coupon Promotion Handling in Retail.

ISBN 978-3-8325-1347-4   40.50 €
270 Seiten, 2006

Bd. 29: Blasius Lofi Dewanto

Anwendungsentwicklung mit Model Driven Architecture – dargestellt anhand vollständiger Finanzpläne.

ISBN 978-3-8325-1480-8   40.50 €
276 Seiten, 2007

Bd. 30: Michael Thygs

Referenzdatenmodellgestütztes Vorgehen zur Gestaltung von Projektinformationssystemen.

Adaptive Referenzmodellierung im Projektmanagement

ISBN 978-3-8325-1517-1   40.50 €
287 Seiten, 2007

Bd. 31: Markus Löcker

Integration der Prozesskostenrechnung in ein ganzheitliches Prozess- und Kostenmanagement.

ISBN 978-3-8325-1526-3   40.50 €
267 Seiten, 2007

Bd. 32: Lev Vilkov

Prozessorientierte   Wirtschaftlichkeitsanalyse   von RFID-SystemenIntegration.

ISBN 978-3-8325-1578-2   40.50 €
253 Seiten, 2007